Poultry As A Meat Supply
Hints to Hen Wives – How To Manage Poultry Economically and Profitably

by William P. Nimmo

with an introduction by Jackson Chambers

IMPORTANT NOTE & DISCLAIMER

IMPORTANT NOTE :

As with all reprinted books of this age that are intended to perfectly reproduce the original edition, considerable pains and effort had to be undertaken to correct fading and sometimes outright damage to existing proofs of this title. At times, this task can be quite monumental, requiring an almost total rebuilding of some pages from digital proofs of multiple copies. Despite this, imperfections still sometimes exist in the final proof and may detract slightly from the visual appearance of the text.

DISCLAIMER :

Due to the age of this book, some methods or practices may have been deemed unsafe or unacceptable in the interim years. In utilizing the information herein, you do so at your own risk. We republish antiquarian books with no judgment or revisionism, solely for their historical and cultural importance, and for educational purposes.

Self Reliance Books

Get more historic titles on animal and stock breeding, gardening and old fashioned skills by visiting us at:

http://selfreliancebooks.blogspot.com/

INTRODUCTION

I am very pleased to present to you another important book on raising poultry – **Poultry As A Meat Supply**. It was written by William P. Nimmo, and was first published in 1916, making it more than a century old. And although an antiquarian text now, it still contains basic poultry management information just as relevant today as it was all those decades ago.

We strive to publish as many poultry titles as we can – they are our best selling line of old book reprints – as more and more people turn away from the dangers of our broken commercial food-chain. We don't want food that's a risk every time we eat it – unsanitary meat or eggs from unhealthy, over-vaccinated, nutritionally-depleted animals fed GMO feed that are not cared for properly.

We want to feed our families good, wholesome, healthy, *clean* food! And to achieve this, it means either spending thousands of dollars more per year on organic, small-farm-raised produce, meat and eggs, or growing, rearing and producing them ourselves.

Features chapters on *The Poultry House, Different Kinds of Poultry, Feeding and General Management of Poultry, Hatching and Rearing of Chicks, Turkeys, Geese and Ducks*, plus others.

This book is a great place to start to learn the fundamentals of raising poultry for meat, and a good book to dive into for beginners, or those even just considering taking the plunge into raising their own poultry for meat.

Jackson Chambers

State of Jefferson, December 2017

Ludlow 1898
The Feathered World

DUCKWING.

PILE.

BLACK-BREASTED RED.

BIRCHEN.

BROWN-BREASTED RED.

GAME BANTAMS.

(Specially drawn to illustrate Mr. Proud's article on Bantams.)

Vincent Brooks, Day & Son, Ltd., Lith.

BLACK ROSECOMB.

JAPANESE.

SILVER AND GOLDEN SEBRIGHTS.

RANTAMS.

BRAHMAS.

WHITE ROSECOMB.

BOOTED.

PREFACE.

THE object of the following pages is twofold. First, to give such plain and practical directions to the hen-wife as may enable her, without any previous knowledge of the subject, to begin and carry on the business of the Poultry-yard, and to do it in such a manner as to make it a *profitable* as well as a *pleasant* pursuit. Secondly, to place the subject in such a light as to induce the rural population to attempt the rearing of Poultry on a larger scale than has yet been done in this country. To give this impetus to the work, the writer has adduced various reasons bearing on the subject in general; while those arising from the particular state of the country in connection with the Cattle Plague have not been forgotten. She can only beg the reader to examine these with candour, and to peruse the work with patience. The result she leaves with the public.

CONTENTS.

CHAPTER I.

CONTENTS.

CHAPTER VI.

CHAPTER VII.

CHAPTER VIII.

POULTRY AS A MEAT SUPPLY.

———◆———

CHAPTER I.

INTRODUCTORY.

TO increase the resources of an increasing population is at all times a praiseworthy object. It is doubly so at this time, when a deadly and infectious disease is making such ravages among our flocks and herds. Already nearly one hundred thousand head of cattle have been swept away; and, should the disease not be speedily checked, it is said that by the month of June there will not be an animal left to die of *Rinderpest*. We, for our part, have no idea that such a gloomy foreboding will be realised; nevertheless, things look sombre enough to call for serious consideration. To the scientific we leave the duty of discovering means of *protection* or of *cure*; to the public authorities that of adopting stringent measures

A

to prevent the spread of so fatal a disease; to the speculative we leave the duty of importing supplies from other countries. Hitherto some of these projects have not met with the encouragement that was expected or deserved; but the scarcity, we must expect, may possibly have the effect of making that a success which was once a failure. Without enlarging on the different products which might with propriety be introduced, we leave importation from any and every quarter to others. The duty we take upon ourselves is to endeavour to increase our home products, so as in some measure to compensate the loss which we have sustained. And passing over all the larger live stock,—leaving the cattle-shed, the sheep-walk, the piggery even,—we settle down in what is considered the least important of all, viz., the *Poultry-yard*. Here we mean to concentrate our thoughts for the present, and to this subject we beg attention. And let not the reader smile to hear the *fowl* named in the same breath with the *ox*, as if the one could be any compensation for the other. Let him think of the Scotch proverb, " Many littles make a muckle." Think of the tiny rill that trickles down the mountain side—hundreds of these combine to form a stream; the stream flows into the river, and the river swells the mighty ocean. So of the domestic fowls—a few of these tiny creatures (tiny in comparison to the ox) would not serve the end we propose; but were they increased as they might be, the case would be far different. Were the units owned by the cottager to be increased to tens, were the tens of the country-house to rise to hundreds, and were the hundreds of the farm-steading to mount to thousands, surely this would be no mean compensation for our recent losses, and would tell vastly on the health and comfort of the community at large. At present the farmers, urged by fear of the plague, are pushing their fat cattle into the market with all possible speed, so that, far from there being any want of animals, the market is quite glutted, and prices ought therefore to be low, though we know too well they are far from being so. But in a few months this will all be over; the supply will gradually be exhausted; then will the effects of the epidemic be felt in reality, and prices, it is to be feared, will rise in a manner of which we have now no idea.

It is desirable, then, that at this season there should be an abundant supply of poultry to bring into the market; and we do not see why it should not be so. In France and Belgium, where immense

quantities of fowls and eggs are consumed, they not
only produce sufficient for home consumption, but
annually export large quantities. We in Britain, on
the contrary, are compelled to import largely for the
supply of our own wants; and not only so, there is a
steady increase in the quantities imported. In 1848
the number of eggs brought into the country was not
much above seventy millions; in 1861 it amounted to
upwards of two hundred millions. The returns for
poultry have not been made since 1856; but, taking
the average number and increase for some years previous to that date, it would make for 1861 an amount
of fully £360,000 for poultry and eggs. But this is a
very low computation—we should say that half a mil-
lion was nearer the true value of the importation for
1861. Doubtless, since that date, the increase has
been going on, perhaps, with accelerated speed, keeping pace with the growing population, so that we are
increasingly dependent on our neighbours for what
might with great ease be produced at home. And
is it not a pity that such a large sum should find
its way into the pockets of foreigners, instead of
those of our own population?

That our countrymen would find it greatly to their
advantage to cultivate more carefully this branch of

rural economy, there cannot be a doubt. A stimulus has of late been given to it, but this has been confined principally to amateurs, and has, as yet, had little influence on agriculturists in general. To these last it must extend, ere it can have the desired effect of fully supplying our markets with fowls and their produce, and thus lessening our imports. The poultry-fanciers have indeed done much : they have vastly improved our breeds, and have introduced numerous and beautiful varieties. By their public shows and competitions they have given an impetus to this branch of rural economy, the benefit of which we are just beginning to reap. Nevertheless, it is on too limited a scale. Few, comparatively, have either the *purse* or the *patience* to carry on such a work. We want something broader, more general; we want the whole rural population to engage in it—every cottage with a kail-yard we would have enlivened by poultry; around every house, which could boast of a garden, we would like to see our feathered favourites; but we would have them for *use*, rather than for *ornament*. Nevertheless, we do not exclude pleasure; nay, we cannot do so, for in such an occupation there is a pleasure which only those who have been engaged in it can comprehend. Even the care of a dozen barn-

door fowls involves a happiness and delight which
the listless daughters of indolence may well envy;
and, to the true lover of nature, there is even in
the plumage of the most ordinary hen a beauty which
never fails to call forth admiration. And in the
rearing of the commonest brood, what simple delight,
what charming excitement, pure and healthful indeed, but not the less stirring. How pleasant to
choose a tried and steady mother-hen, to watch and
tend her while she patiently performs her apparently
wearisome task! And when the time arrives for the
imprisoned chicks to burst their shelly fetters, how
delightful to reckon them up one by one; and should
there be the full complement, how triumphantly does
the hen-wife tell of her success. For if she is worthy
of the name, she enters heartily into her work, and
her happiness is, in some measure, bound up with
her success. But to enumerate the charms of this
department of rural life were needless, and would
occupy too much space; suffice it to say, they are
many and varied, ever fresh and ever new. And we
do not see why they should not be enjoyed by every
one whose privilege it is to live in the country,
among green fields and pure streams. Pleasure and
profit, we are persuaded, would go hand in hand,

were the work gone heartily into; and in view of the enormous sum spent yearly on poultry, surely no one can say it is a paltry pursuit.

The value of eggs annually consumed in England is estimated at £3,000,000 sterling. When we add to this the millions of fowls annually consumed, we see that is a matter of considerable importance, and worthy of ranking among the objects which may engage the attention of our first-class agriculturists. Were they in a body to put their shoulder to the wheel, our object would speedily be accomplished: not only would the large imports above mentioned be unnecessary, but the markets would be filled—not glutted, however; for the scarcity which by mid-summer we may reasonably expect, will open the way for the ready sale of all that the agriculturist can produce. It is to be regretted that every species of domesticated fowl has been so much neglected by the farmer. While he spares no expense in the housing, and tending and feeding of his four-footed live stock, he merely tolerates the hapless bipeds of the feathered tribes, and seems to reckon every bushel of grain bestowed upon them as utterly wasted. The cry has been, "They do not pay;" but now it is pretty generally acknowledged, that if they are unprofitable, they are only so by mismanagement, or rather by no management at all. If they are allowed to forage for themselves, to pick up a livelihood as best they may, how is it possible that they can fatten so as to bring a price in the market, or even to make a respectable appearance on the farmer's table? Let cattle, sheep, or even pigs be treated in a similar manner, and what would be the result?—we trow they would be equally unprofitable. In France the *basse-cour* of the farmer has no mean share in making up accounts on rent-day; and in our own country, not a few first-class agriculturists rear poultry on a very extensive scale; and it is said that, notwithstanding the costly appliances provided for them, the profits they yield go far towards meeting the landlord's claim. Be this as it may, we are persuaded that, by following the proper method, poultry will pay, and pay, too, a fair, nay, a high percentage. But then they must stand on equal ground with the other denizens of the farm; must have a fair share of attention; be well tended, well housed, and well fed. The mason and the joiner must not be grudged for them any more than for the cattle. They must not be put off with the refuse of the farm, bestowed one day and withheld the next. A fowl will not fatten,

any more than any other animal, by feasting and
starving alternately. The corn must not only be
supplied to it *liberally*, but *steadily*. Were the farmer to take some trouble in providing accommodation
for his poultry, and to bestow supplies with ungrudg-
ing liberality, he would find it much to his advantage.
The ladies of his household, meeting with such encouragement, would bestir themselves more, and take
a pleasure in seeing their yards ornamented with
fowls the best of their kind, and that kind the best
adapted for the climate and situation. Until this is
the case, little will be done; but when the ladies of
the farm, and of rural districts in general, take up the
matter in good earnest, a vast change will speedily
be effected. Those lean, bony, skinny, hungry-look-
ing creatures that too often disfigure the homestead
will disappear, and their places will be supplied by
plump and beautiful birds of every variety, numerous
enough plentifully to supply the family and the
market; and the farmer, when he finds his butcher's
bill steadily diminishing, and his coffers filling in a
manner he did not anticipate, far from grumbling as
was his wont, will congratulate himself on the agreeable change, and acknowledge his error in joining the
general cry against these much-maligned creatures.

CHAPTER II.

THE POULTRY-HOUSE, ITS PERCHES, NESTS, LADDERS.—
THE YARD, ITS FEEDING-VESSELS, WATER-TROUGH.
AND DUST-BATH.

ASSUMING the fact, then, that, if managed properly, poultry will pay, we proceed to point out in detail the method which we conceive fitted to produce that result. And before proceeding to enlarge on the feeding and general management necessary, a few words regarding their houses and yards may not be out of place. Amateurs, whose object it is to improve the breeds in every possible way,—to produce birds of rare beauty, and who must therefore keep each variety separate from the other,—do not grudge expense while providing accommodation fitted to obtain these results. They have their suites of apartments (if we may use the expression), their yards and their runs, and all the various appliances which wealth can provide. But as these do not enter into our plan, and, instead of promoting our object, would tend to defeat it, we forbear to describe these costly buildings, or to offer any plans for their erection, but content ourselves with referring our amateur readers to the popular work of Mrs F. Blair, where they will find in full detail all that can be desired by the most fastidious fowl-fancier. Our object being to *fill* rather than empty the purses of our readers, we naturally leave costly buildings on one side, and prefer recommending that each and all should take advantage of the accommodation already existing in their respective premises. A little ingenuity and skill will, at a comparatively trifling expense, make comfortable quarters for fowls out of the various erections which are to be found connected with every country residence, and which are frequently employed merely for the reception of useless rubbish. And then, in the farm-steading, there is found invariably the hen-house—not always, indeed, in the best situation, or in the best state of repair; but there it is, and so let us make the best of it. And first let the farmer see that it be completely weather-tight, thoroughly impervious to rain and snow, and sufficient to exclude the biting blasts of the wintry air. Let him not grudge the workmen's bills for his poultry-house, any more than he does for his byres and his stables. Let him see, also, that it is properly lighted and well ventilated. If the exposure is south, so much the better; and a door to the east will give the benefit of the morning sun. Light in the poultry-house is an absolute necessity; the inmates must have it, were it for nothing more than to see their way to their nests and roosting-places. For health, also, it is indispensable; animals, we fancy, will not thrive in the dark, any more than plants. And should any object to this, and say, " Why, then, are fowls placed in dark cellars when being fattened for market?" Our answer is at hand :—Fowls when crammed *do* attain to an extraordinary size and fatness, but it is not a *healthy* fatness, and borders so very closely on disease that, were the process continued only a very little longer than the prescribed time, the crammed fowl, be it turkey or hen, would be rendered totally unfit for human food. And if light is necessary, air is equally so. This we learned by experience, an experience of rather a vexatious nature. Being totally ignorant of the management of fowls when we commenced the labours of a hen-wife, and being scantily provided with books for our guidance, we made a notable mistake on that score. Requiring additional accommodation for our chickens, we fixed upon an out-house which seemed to possess every requisite. It looked to the east, thus admitting the morning sun; it was well lighted, thus giving the chicks an opportunity of picking as soon as day dawned. Knowing this to be of *immense* advantage in rearing them, we expected *immense* success. Instead of this, what was our disappointment to see an unusual mortality prevail. The cause we could not imagine, and only discovered it by the accidental visit of a lady from the country, well skilled in such matters. We took her to see our "treasures," among others the remaining inmates of the new chicken-house. Scarcely had the door been opened ere she exclaimed, "No wonder than your chickens die; the want of air here is sufficient to kill twenty chickens in one night." On examination we found that the chimney, which was sufficiently wide to admit the air required for the size of the place, had been built up at the top or roof, the windows were quite tight, and there was no little square opening in the door, as is usual in regular hen-houses. There was therefore no possible inlet for air; and had we not generally during the day kept the door wide open, doubtless in one night all would have perished. Fortunately our loss was not so great as this; however, it gave us a lesson which was of use in after years.

In winter, especially during the time of a storm, every precaution must be taken to protect the poultry from cold. The doors and windows must be carefully closed, and the roof may be covered with straw. Snow forms a warm covering for the roof of the hen-house; but the warmth arising from the assembled inmates is apt to melt it, and, as is well known, snow, being of a searching nature, will probably find access to the hen-house, and make it wet and damp, than which nothing can be worse for fowls. To avoid this the snow may be removed, and be replaced by straw; the snow falling on this will form a warm and dry covering. Should the hen-houses be heated by flues, care must be taken to regulate them properly. While fowls, being originally from warm countries, suffer greatly from cold, they do not thrive with too much heat, but do best in a moderate temperature. If the hen-house be in the vicinity of the byre, or stables, or boiler, the warmth thereby obtained will be found very advantageous. In some homesteads we have seen poultry of various kinds roosting on the "joists" in the byre, and can testify that they throve à merveille. Even the cottage need not be without its warm hen-house. A small erection may be made just behind the fire-place, three sides only being required; it could be put up at little expense, and having the advantage of the adjoining fire, it might be of wood, provided it were made firm and water-tight. Another plan to utilise both heat and space in rural cottages has been proposed, viz., to fit up the space usually left between the roof and the flooring with a few perches for roosting, and a few nests and nest-eggs; to admit the fowls, a small opening may be made in the wall, which may be reached by a light hen-ladder. For the purpose of cleaning, etc., a trap-door may be made in the most convenient part of the house. The space nearest the fire may be used first and extended at pleasure. In such quarters a few broods of chickens might be successfully reared *in winter*, and, being brought early into the market, would command a high price. They would thus soon reimburse the cottager for his outlay, and not only so, but would add no inconsiderable item to his weekly income.

When the want of accommodation necessitates the building of a new poultry-house, it ought to be of stone or brick, wood being, generally speaking, unsuitable for this variable climate. A southern exposure is the best; failing that, the east may be chosen, but the north must if possible be avoided. A yard must be provided adjoining the hen-house, in which the fowls may feed and scrape about during the day; and were half of it laid down in grass, and the fowls turned out into it each alternate day, it would be found very conducive to their health. But this applies only to situations where space is limited, not to the farm-steading, where for so many months in the year the fowls have the run of the stack-yard, stubble and grass fields, and therefore do not require any such enclosure. It has been well said that " a farm-house is the paradise of poultry," and truly the liberty and plenty there enjoyed is of more value than all the appliances that wealth can give.

But to return to the hen-house. The floor should be well laid, covered with a composition of sand-lime and chalk, thoroughly mixed and beaten hard; it should slope slightly towards the door, and be raised a little above the level of the yard, so that it may be kept dry and be easily cleaned. The roosting-perches should be fixed in the wall about a foot apart; if more than one row be required, they must not be placed directly under each other, but in a sloping direction; they should be rough and of a square form,

B

so that the fowls may grasp them comfortably; they cannot do so when the perches are smooth and round. Nests for laying must also be placed in the hen-house, provided with plenty of clean straw, and large chalk eggs to induce the hens to lay. As the hen is very prudish where laying is concerned, care must be taken to put the nests in the darkest part of the hen-house; they must be raised a little from the ground, to protect the hens from cold draughts, and also to secure them from damp. To increase still more the privacy, the nests may be boarded up a little way; they may be placed in tiers one upon another, in the form of square boxes. Some prefer small oval baskets hung on the wall. In a large hen-house both plans may be adopted.

The roosts should not be too high, as the fowls are apt to injure their feet in flying down, for though they ascend by means of the ladder, they never by any chance use it in descending. Turkeys roost very high, and this is quite safe in their natural state, but far from being so in a small crowded hen-house, with a hard floor to light upon. In descending from trees, which is their natural roosting-place, they have abundant room, and soft grass for a landing-place. The natural propensity of the turkey in this respect must be restrained, if we would protect it from injury. Ducks and geese roost on the ground, and should have a separate house. Considering the numerous out-houses, sheds, etc., which are to be found in a farm-steading, this will easily be obtained. A hatching-house is also necessary, for hens will not sit if disturbed by the cackling noise of laying hens. Repose and quietude are absolutely necessary. If there are many sitting at a time, they must be separated, so that one cannot steal the eggs of the other; this they are sure to do if they have the chance, and the adroitness with which they contrive to convey the eggs from one nest to another is truly surprising. The larger and whiter the eggs, the greater the likelihood of their being stolen. We say nothing of the size of the hen-house, as that depends so much on the number of fowls kept. In summer they require more room; over-crowding in the heat of summer will induce disease, but in winter they may with safety be more closely packed: in frost and snow a well filled hen-house gives an idea of warmth and comfort. We have frequently had thirty full-grown fowls, besides chickens, crowded into a place little more than ten or twelve feet square, and found it do very well. In summer, however, we had to exercise our ingenuity

in adapting our premises to our increased require
ments. The hen-wife will easily perceive if her ac
commodation be deficient, and the farmer will consul
his own interest by promptly supplying the wan
There is no reason why money expended on goo
housing for his poultry should not bring a returr
as well as that laid out on his other live stock
and, other things being attended to, it will doubtles
do so.

In addition to warmth, cleanliness must be attende
to; the floors of the hen-house ought to be swep
every day, and the perches occasionally washed witl
clean water, in which a little lime has been mixed
The whole hen-house,—walls, roofs, perches, and nests
—should undergo a thorough purification, at least twice
a-year. The floor, after being swept, should have
sand, ashes, or gravel scattered thickly over it. Some
think saw-dust preferable to all, as being more cleanly
for our part we prefer the three first-mentioned mate
rials, any or all, as may be convenient; the cinder
and small stones they contain assist digestion in the
fowl—indeed, they are absolutely necessary to its
well-being.

A heap of dry sand or ashes should also be placed
in the hen-house or yard, at some place to which the

owls have constant access. They require it to cleanse their feathers, and to free them from the vermin to which, even under the best treatment, they are subject. It may be mixed with any limy rubbish that may be at hand, or if that cannot be had, oyster-shells, burned in the fire and then pounded down, will answer the purpose. The hens, while performing their dry ablutions, if we may be allowed the expression, will pick out the materials for grinding their food, and also the calcareous matter necessary for the formation of egg-shells. The heap of ashes, called the " dust-bath," ought to be placed under cover to protect it from rain and to keep it dry. Considering the enjoyable glee and spirit with which a hen goes through the process of " dust-washing," so to call it, any negligence on the part of the hen-wife, tending to abridge or lessen this pleasure, is quite inexcusable. The next thing to be attended to in the furniture of the yard, is the feeding-vessels. These are usually flat dishes or trays. The fowls invariably tread upon them with their feet, thus dirtying the food; to prevent this we have seen narrow strips of wood or gutta-percha placed on the trays diagonally. The fowls tread on these strips and pick out the food at the openings. Being also prevented by these diagonal fittings from scraping, as is their wont, the food is kept partially clean. Perhaps the best feeding-vessel is a long open trough, similar to that used for pigs, but smaller. To prevent the fowls upsetting or dirtying it, it may be placed behind iron rails on one side of the yard. The rails may be driven into the ground about three inches apart; they should be all of one height, thus admitting of a sloping wooden roof being put on, which will protect the food from rain. It is a very convenient plan to have a covered shed in the yard, at one end of which may be placed the dust-bath, and in the other the feeding-trays. But when the weather is dry and the yard is kept clean, well covered with sand and gravel, nothing can be better than merely scattering the food on the ground; it ensures a more equal division of the food, and the sand and small stones picked up with it assist in its digestion.

When the yard is furnished with a drinking-vessel, its fittings will be complete. This may be the same as the feeding-trough, and protected in a similar manner. It, as well as the feeding-vessels, must be daily and thoroughly cleaned, for nothing is more indispensable to the prosperity of the poultry-yard than the most scrupulous cleanliness. Care must be taken

also to have a constant supply of pure fresh water—
a condition of prosperity quite indispensable, but,
happily, at the same time, so easily attained that the
neglect of it is altogether inexcusable.

CHAPTER III.

THE OCCUPANTS OF THE POULTRY-HOUSE.—HOW TO CHOOSE THEM.—DIFFERENT KINDS OF POULTRY— THE DORKING — SPANISH — POLISH — HAMBURGH — COCHIN—CREVE-CŒUR—GAME FOWL.

HAVING now provided our hen-house, and seen it properly furnished with its perches, its nests, and its ladders,—having seen our court-yard provided with its dust-bath, its feeding-trays, and its water-troughs,—having prepared our grass-run, in which our feathered favourites may take their daily airing,—having done all this—provided house and grounds—let us choose the occupants of the domain. In other words, having provided the poultry-house, let us choose the poultry; and, be it remembered, this is a matter of no small consequence,—a mistake here will materially affect the future prosperity of the poultry-yard. Let the hen-wife, then, be on her guard as to the fowls with which she begins.

Condition here is of more importance than *kind*,
at least to the hen-wife, whose desire it is to have
profitable poultry. The very best kinds may be
seen every day scraping about our streets, and picking up a scanty livelihood as best they may; but the
slightest observation may convince any one that such
ill-fed, ill-conditioned creatures, could never produce
chickens calculated to be in any way profitable to
their owners. The breed may, indeed, be pure—
Dorking, Spanish, or Polish, as the case may be; but
the barn-door fowl, if from its earliest days it has
been well tended and well fed, will be found infin-
itely better for the hen-wife, who would have her
poultry pay, than those half-starved accumulations of
bones and feathers, however pure the kind may be.
The face of the Spanish may be white according to
the studied rules, the five toes of the Dorking fully
developed, the crest of the Polish equally so; but
what matters it? these points, and twenty others
added thereto, would not make up for the woeful condition to which neglect has reduced these fowls, and
which is sure to be perpetuated in their progeny.
Whatever be the kind fixed upon, let them be in
good condition, and chosen from a yard where
care and attention have been bestowed upon them

from the time they emerged from the shell, and on-wards. The success we had in our poultry-yard we attribute in no small degree to the excellence of the stock with which we commenced. It consisted of two Spanish and four Dorking hens, and a splendid, sprightly cock, with beautiful bright plumage. All were in good, first-rate condition—young, large, fat, and lively. The hens began to lay in November, and continued all through the winter. On an average they laid four eggs a-day. This the hen-wife must consider good; if she expects an egg a-day from each hen, as in our ignorance we did at first, she will certainly be disappointed.

Well, to return to our hens; we being at first rather timid, as was to be expected, only attempted one or two sittings the first season. Emboldened by success, the next season we speculated more largely, and with increasing success, until in a few seasons our stock of seven had increased to somewhere about a hundred; and had we set every hen that showed unmistakable signs of her desire to perform the duties of maternity, our number would have been more than doubled. Fain would we have done so, for it would have been pleasanter by far to indulge their natural instincts than to run counter to them—instincts which we

frequently found so strong, that days of solitary confinement could not extinguish them; even the cruel
method of ducking we often found ineffectual; and
only after sitting for weeks upon an empty nest,
would the poor, disappointed thing give up the attempt as hopeless.   We have often regretted that we
had not given way to the instincts so strongly developed in fowls, and as often speculated as to the
increase of which we might have boasted in a given
number of years.   And we are assured that, had we
followed this plan, our stock would soon have increased tenfold.   Some extra expense as to accommodation would have been incurred, but that, as well as
the expenditure for additional food, would soon have
been returned with interest.   And we strongly recommend those who reside in the country, and who
have premises and ground that might be turned to
account, to try this plan, at least for a time, and to
allow their stock to increase as instinct directs—we
are sure they will find it pay.   The sitting-hen con-
sumes but little food.   The chicken, for the first two
months of its life, costs but little; the next two it
will cost more, but, if properly managed, it will then
be fit for table or for market, and will be worth
nearly double the sum it has cost.   The great thing

is to feed *regularly*, and the moment it is ready for market, despatch it thither without delay; every grain it eats after this is just so much thrown to the winds. And this putting off, this allowing of the fowl to live after it is ripe, is one fruitful cause of the popular cry, " Poultry will not pay." How can they pay if the chicken is allowed—if we may use the popular phrase—to " eat its head off" time after time ? But we are anticipating, and must return to our proper subject.

Finding that the eggs produced in our own yard uniformly came out best, we used them chiefly for setting. Occasionally we got a hatching from a totally different strain—sometimes pure Dorking, sometimes Spaniards; common barn-door fowls, even, we did not despise. Our beautiful cock, having been lord of the dunghill beyond the prescribed term (two years) had to give place to another, or rather to others; for now a numerous train of chanticleers had to fight for supremacy, and many a bloody battle was fought ere the victory was won.

By this plan our stock became remarkable for the variety, if not for the purity of the breeds. We had Dorkings, pure and mixed—Spanish, too, of both kinds—beautiful grey merlins—common hens, of rare

beauty, that vied with the pheasant and the partridge
—others with a dash of the game, a few with the
Cochin predominating.   In short, we had a mixture
of every kind, and all in first-rate condition, ready for
the table at a moment's notice.   This is an advantage which the housewife will be able fully to ap-
preciate, especially if the wavering appetite of a sickly
child demands more delicate food than the beef and
mutton which hitherto have formed our staple com-
modities.   And should she reside at a distance from
market, and her friends take it into their heads to
visit her at a time when her larder is but ill supplied,
the convenience of being able to furnish her table
from her own yard will be felt in a manner which
those who reside in the heart of a town, in the vicinity
of marts and markets, cannot imagine.   And is it not
a pleasure to the housewife to present to her family
and to her friends fowls that she knows must be tender and good, ham which she can testify has been
well fed and as well cured, and vegetables freshly
gathered from the well-tended garden?   Yes, truly it
is; and that her own labours have contributed to this
result is no small addition to the pleasure.   But it
may be said that we are diverging from the subject.
True, in some measure; yet not altogether so, for we

are convinced that the housewife who devotes her attention successfully to the rearing and tending of poultry will not stop there, but will extend her labours to other branches of rural economy, and doubtless with similar success. But leaving these branches to others, we return again to our own peculiar province.

Hitherto we have advocated the mixed breeds, where large numbers are kept, for several reasons; and, in the first place, from the great difficulty there is, in these circumstances, of keeping the breed pure. Where you have only six or eight hens the thing is easy, but it is quite different when your number mounts up to hundreds. You may begin, indeed, with the purest and finest specimens, but the chances against their continuing so are very strong. Then, if you have separate apartments for your " aristocrats," so to speak, you must necessarily deprive them of the liberty enjoyed by your " commoners;" and thus you will in all probability soon find that, deprived of this privilege, they will not thrive. Again, the keeping of the several varieties apart involves an amount of labour and expense which would go far to frustrate the object which we propose in this work, viz., to make the rearing of poultry a profitable as well as a pleasant pursuit. For the benefit, however, of those who are willing to incur the expense, and to expend the labour inseparable from the rearing and preserving the finest kinds, we conclude this chapter with a description of some of the most approved varieties, and with a few hints as to the management likely to be successful. And the first we mention, as pre-eminently suited to our climate, is the.

DORKING.

As its name implies, it is a genuine old English fowl from Dorking, a town in Surrey. Brought over, it is said, by the Romans, it has had abundance of time to be naturalised. Its distinguishing characteristic is its having five toes on each foot, one of which, however, is often imperfect. Its body is large and handsome, and its legs short. It is of various colours —originally white, but now both grey and red varieties are common. The grey colour is the most esteemed. The flesh of the Dorking is not so white as that of the common hen; but it has a fine flavour, and, from its being so easily fattened, it forms one of the best table birds. Its eggs are very large, and tolerably abundant. It is a good mother, sits early, and thus is invaluable in bringing out early birds. It is frequently used for hatching the eggs of the Spanish, Polish, and Hamburgh. It thrives best in a dry soil, and does not do with damp, clayey ground. The chickens are difficult to rear—so much so, that it is said that not more than two-thirds of a brood survive their fourth week. The Dorking is a great wanderer; no fence under seven feet high will keep it within bounds. The plumage of the male bird is beautifully varied.

Next in order we would mention the

SPANISH FOWL.

This, too, has been long in England, and has been brought, not from Spain, as its name implies, but from India. There are three varieties—black, white, and blue, or Andalusian. Of these the black is the best known, and also the most beautiful. Its white face and ear-lobes, and bright scarlet wattles, form a fine contrast to its glossy black plumage—a plumage the beauty of which is heightened by the rich green and purple tints which it reflects. Nor is the beauty of the Spanish fowl its only advantage: it lays eggs of an unusually large size and in great numbers, and continues to do so during most part of the year, and

thus merits the name it has gained of Everlasting
Layer. Its flesh has a fine flavour and a beautiful
whiteness. The only drawback it has as a table bird
is its black legs, which certainly are no ornament. It
is more domestic than the Dorking, and will not
wander much if well fed. The hen seldom shows
any desire to sit; but this we consider a good point,
as it gives her more time to produce eggs, and generally there are abundance of broody hens of other
kinds ready to hatch them. The chickens are tolerably hardy after they are feathered, but require extra
care until that time. They feather but slowly, and
have a very naked and shabby appearance for the
first six weeks of their life.

The next we mention is

THE POLISH.

These fowls are distinguished by a crest or top-
knot of feathers, the colour of which varies in the
different kinds. The black bird has a white crest;
that of the golden-coloured is of a dark hue. Their
legs are short and bodies plump. They yield in
beauty of appearance only to the game fowl. They
evince very little disposition to sit, and therefore produce a great number of eggs. On this account they

have been called "Every-day Hens," and are considered the most profitable of all varieties. They are, however, delicate when young. Common fowls with crests show a descent from the Polish, and are usually good layers.

SILVER HAMBURGHS

are known by several names — "Bolton Grey," "Dutch," "Every-day Layers." They are excellent little birds; and, what is of consequence to the cottager, are easily kept. As their name implies, they produce eggs in great abundance, though it must be confessed they are very small. The Golden Hamburgh is more uncommon, and very beautiful, but it comes under the denomination of a fancy bird, rather than under that of a profitable one.

THE COCHIN

must not be omitted in our description. Since the "Cochin fever" abated, these birds have ceased to be favourites; but this is only the reaction consequent on the ridiculously high place which was at first assigned them; when this ceases, we are assured they will rise to their proper level, and be valued as they ought. Where one kind only is kept, and the accommodation limited, better could not be chosen, for they thrive in a small space; they do not wander, even if they have the chance; then they are too large and heavy to fly over even a low enclosure. In the next place, they lay far on in the season, and are easily fattened for the table, any want of delicacy in their flesh being compensated by the additional quantity supplied. The frequent desire of the hen for incubation is her great fault; but in a poultry-yard composed chiefly of Spanish and Polish fowls, this would be an advantage; her time would not be lost while hatching the eggs of those " Everlasting Layers." The first cross from a Cochin may be good, but the second, according to high authority, is worthless. In general, we believe that most kinds of fowls are deteriorated by having any admixture of the Cochin; we therefore counsel for them entire isolation. A few pure white Cochin hens enliven and beautify a poultry-yard greatly; but we would carefully exclude the male bird, for the reason above mentioned.

THE CRÈVE-CŒUR.

A French bird, which has lately been brought to this country, and merits our high commendation.

We cannot agree with those who think it a "hob-goblin-looking creature," "very like a devil," for, in our opinion, it possesses beauty quite sufficient. The hen is, indeed, a heavy-looking bird, but this is owing to her great size; the shortness of her legs, too, makes her look almost as if she were crawling on the ground, but her beautiful crest gives her a noble appearance, and her pendent cravat invests her with a matronly air. The cock is, unquestionably, a splendid bird; his glossy black plumage reflects beautiful green tints; he has a sprightly majestic air; and his feathery cravat, his cloven comb, and his tufted head, add not a little to his beauty. The Crève-cœur does even better than the Cochin in a limited space; she prefers the dunghill to the largest run, and seems afraid to move from her accustomed retreat; she is very timid, and of great "amiability;" she scrapes but little, and notwithstanding her great size, requires but little food; her flesh is very delicate, and she acquires both flesh and fat with a rapidity unknown to any other breed of fowl; she is a first-rate layer, and rarely sits. It is to be hoped that a bird, in every respect so valuable, will soon be largely propagated in this country.

The last we shall mention is the

GAME FOWL.

This bird, although happily not now used for the purpose its name implies, is still in high repute for its elegant and compact appearance, as well as for its many excellent qualities. As good layers of delicious eggs, as excellent mothers and rearers of chickens, game hens have no superior. They are hardy in constitution, and excellent caterers for themselves and for their progeny. The chickens feather rapidly, and with ordinary care and liberal feeding, are as easily reared as those of other fowls. For delicacy of flavour, they are unsurpassed. The disposition to fight is the only drawback to this " prince of breeds."

The hen-wife, having now before her the most approved kinds of poultry, with their good and bad points, is in a position to choose her own stock. This, we confess, is a difficult matter; as in other things " wealth makes wit waver," so in the choice of fowls. When all are so good, it is hard to leave out any, still harder to be limited to one; and so, possibly, the hen-wife may wish to try several varieties. In all probability she may choose a pair of prize Dorkings, from which she may propose to breed; another, perhaps, of pure Cochin; and a third, of fine French fowls. She must, of course, have separate apartments for each kind, possibly on a small scale; it matters not, for much room is not required. Let her see, however, that, in addition to constant pure air, her fowls have also the benefit of sunshine; this, in winter especially, is of great consequence. Let not the sprightly, roving birds be placed in adjoining pens, lest they should fight through their fences, or invade each other's domain, but put the heavy fowls, such as the Cochins, in the centre pen, between two rovers. Being placed in pairs is almost solitary confinement to poultry, for their nature is to go in flocks. Let the hen-wife then put to each pair two or three hens, the eggs of which can be distinguished by colour, form, or shape, from those she means to breed from. This additional society will put life and spirit into the imprisoned pets, otherwise they will be sure to droop and pine. This we learned from experience, that peerless teacher, whose teachings, unhappily, are often too late to be of use.

Should the hen-wife decline the toil of separate varieties, and desire her general stock to consist of one kind, and that to be pure, perhaps the Dorkings will, on the whole, be found the most suitable. They combine all the qualifications necessary to constitute a good and profitable fowl. They are good foragers, good layers, good sitters, good mothers, and good for the table. The Spanish fowl is to be preferred before all others, when the principal object is a large supply of fine eggs, for she produces them in great abundance, seldom showing any desire to sit. The eggs are also very large; but the hens are bad foragers, bad mothers, and hard moulters. A cross with the Dorking overcomes all these bad qualities, retains the good, and produces a fowl which for laying and breeding cannot be surpassed, and which for the table may challenge anything that can be produced.

In choosing fowls, much depends on the nature of the soil and climate. Low-set fowls, such as the Dorking, do not thrive on damp clayey soils, while the long legs of some other kinds serve as stilts to protect them from the moisture. But whatever be the breed chosen, let the fowls be the best of their kind—healthy, handsome, full-bodied hens—and let the cock be robust and sprightly. As to the number of cocks required, one to six or eight hens is usually allowed, if prolific eggs are wanted; otherwise, one for a yard of fifteen or twenty hens is sufficient. Let

the cocks be changed every second year, and procured from a totally different strain. Let them be
the very best that can be had for love or money. By
attending to these hints at the outset of her career,
the hen-wife will spare herself much future annoyance,
and make her poultry-yard eventually both pleasant
and profitable.

CHAPTER IV.

FEEDING AND GENERAL MANAGEMENT OF POULTRY.

WE have now prepared and furnished our poultry-house, and chosen the inmates destined to occupy it, but let us not suppose on that account that our work is done. Little thanks would we deserve from those depending on us for placing them in a pleasant habitation, however well it might be provided with all that could minister to comfort and convenience, if, at the same time, we neglected their table, or supplied it but scantily. And so our feathered favourites will have but little enjoyment in the most comfortable quarters, though constructed according to approved principles, if their stated supplies are withheld, or given with a stinted and niggardly hand. No; in order to their well-being it is not enough that they be well *housed*, they must also be well *fed;* and the hen-wife will find it to her advantage to be liberal in this respect. As has been so often said,

" If it will not pay to keep them well, assuredly it will not pay to keep them ill." The hen-wife who, on hearing the reiterated statement that poultry *can* and *do* pay, conceives the idea that they can be reared on little or nothing, is labouring under a grievous mistake. That fowls, any more than other live stock, should be *all profit*, is an error too unreasonable to be combated. A *fair* profit is all that is promised, and, we may add, all that a reasonable person will expect; and it may safely be affirmed, that the more liberal the feeding the greater will be the profit. Food, indeed, must be given with judgment and discretion; waste must be avoided on the one hand, as well as parsimony on the other. To strike the proper medium —to be liberal and yet economical—is the object professed in this chapter. Before proceeding further, however, it is necessary to add that these remarks apply to the keeping of fowls in large numbers. In the cottage or the private family, where but few are kept, and where they are fed from the scraps and crumbs that fall from the table, the case is different. Consuming only what would otherwise go to waste (with the exception of a little dry grain, which should never be neglected), they bring to their owners nothing but profit. Of these we are persuaded thousands might be kept in this country. We say thousands; but we believe this is far below the mark, and the following quotation will show that in this view we are not singular:—

"I calculate that, without the expenditure of anything but what would otherwise go to waste, ten millions of common fowls might be constantly kept in Great Britain and Ireland. These might supply twelve millions of chickens and fowls for the table, at the average of fifteenpence each—in the whole, £750,000 sterling. They would afford, also, for the table twenty-four millions of eggs, worth, at one halfpenny each, £50,000 sterling. Thus the annual produce of our common fowls alone might raise provisions to the value of £800,000, from a capital stock of the same value, entirely without expense of labour or food that does not otherwise go to waste. The facts on which I make this statement are undeniable."

Before entering in detail on the different plans of feeding and fattening of poultry, we would premise that whatever be the food, it is of the utmost consequence that it be given *regularly*. The richest feeding, given by fits and starts, will never fatten. It may be abundance to-day and none to-morrow, and scant supply the day following; then perhaps another feast, followed by successive fasts, and so on indefinitely. Such treatment will never fatten; poorer food by far, and less of it too, if given at stated times and regular intervals, will be far more efficacious in making our poultry fit for table or market. The starved fowl, like the starved child, will be characterised by a voracious appetite,—an appetite which, though satisfied to the full, will contribute but slowly to its improved condition. If the hen-wife, then, would study economy, she must consider not only the quality of the food, but the regularity with which it is given. As fowls are early risers, and in summer leave their roosts soon after sunrise, it will be advisable to give them a little light grain between five and six o'clock in the morning. Should this be inconvenient, it could be scattered in their yard or under their shed the night previous. They will have sharpness sufficient to discover it if left uniformly in the same place, for they are far from being stupid, especially with regard to the " supplies." Their regular morning meal may be given some hours later, and may consist of potatoes boiled very soft, beaten up with sharps (sometimes called thirds or middlings), and any pot-liquor or refuse bits of suet and meat which the kitchen may furnish, seasoned with salt. This will be found an excellent mess. Potatoes are the staple food of poultry as of pigs, and their nutritious properties cannot be questioned. The fine bran (sharps) is remarkable for its warmth-giving properties. Notwithstanding all that has been said against it, as being deficient in nutritious properties, it is ascertained that the amount of albumen and oil it contains is 24 per cent. This proportion of nutritive matter is surely sufficient to render it a good ingredient to aid with others in the fattening of poultry. Of course, it would be of little use alone, but it forms an excellent addition to potatoes or any soft food, imparting to the mixture a firm consistency, making it, in short, friable, a property which is of great advantage, as anything glutinous, adhesive—"sticky," we would call it—is exceedingly annoying to fowls. It is also of great use as a medium of conveying nourishment, in the shape of fat broth and rich liquids of different sorts, of which fowls are very fond; and another great advantage is its cheapness. There are three kinds of it; the two finest we can recommend for the purposes above-mentioned, but the coarsest kind is of little use. Since variety as well as quality should be considered, barley-dust may be substituted for sharps, or half barley-dust and half sharps may be used. Another way of varying followed by successive fasts, and so on indefinitely. Such treatment will never fatten ; poorer food by far, and less of it too, if given at stated times and regular intervals, will be far more efficacious in making our poultry fit for table or market. The starved fowl, like the starved child, will be characterised by a voracious appetite,—an appetite which, though satisfied to the full, will contribute but slowly to its improved condition. If the hen-wife, then, would study economy, she must consider not only the quality of the food, but the regularity with which it is given. As fowls are early risers, and in summer leave their roosts soon after sunrise, it will be advisable to give them a little light grain between five and six o'clock in the morning. Should this be inconvenient, it could be scattered in their yard or under their shed the night previous. They will have sharpness sufficient to discover it if left uniformly in the same place, for they are far from being stupid, especially with regard to the " supplies." Their regular morning meal may be given some hours later, and may consist of potatoes boiled very soft, beaten up with sharps (sometimes called thirds or middlings), and any pot-liquor or refuse bits of suet and meat which the kitchen may furnish, seasoned with salt. This will be found an boiling was necessary, and accordingly we recommend this plan to our readers, having found it to suit admirably. As to salt, some have denounced it as poison for fowls; in large quantities it certainly is so, but if merely used as it is for ordinary cookery—a seasoning, as it were—it will be found highly advantageous. According to some it is a preventive of disease, especially the pip and gapes. A slight seasoning of pepper will also be found conducive to health, especially in winter. The mixture thus described should be given moderately warm. If given too hot, it induces disease. It may be either put into the feeding-vessels or scattered about the yard, as previously stated. In addition to their morning mess, the fowls should be supplied with green food in its raw state; cabbages, turnips, beet-root—whatever the garden can produce,—for nothing green comes amiss to fowls. And such food is good for them; they thrive upon it; indeed, it is doubtful that they can do any good without it. Let it then be liberally supplied to them. And last, though not least, pure and fresh water must not be forgotten. The neglect of this indispensable condition of health is the cause of innumerable diseases in the poultry-yard.

With such a morning meal, the fowls ought to be

D

"set up" for the day; and should they be so fortunate as to have their abode in the farm-steading, they will probably require nothing more till roosting-time, when a small quantity of light dry grain may be given them. But should they be in less favourable quarters, they can have towards mid-day a renewal of their morning mess; boiled barley may be used instead of bran and barley-dust, and for economy's sake the hot liquor and minced liver may be omitted. The lazy hen-wife, if such there be, will probably hasten to say, "O, then, we can just put down a double quantity in the morning, and that will save trouble and do just as well." Not so fast, friend, we pray you, for this is exactly the thing you must *not* do. The proper quantity must be given *fresh* at stated hours—not in large quantities, with long intervals between. The sour, and dirty, and clammy state into which such a mode of procedure brings the food, is sure to make the fowls turn from it in disgust; or if they do eat it, it fails to impart the same nourishment as if it were fresh,—either way it is wasted, totally in the first case, partially in the second, and showing clearly that, if the hen-wife would have her "poultry pay," she must not grudge trouble any more than she does expense. Before going to roost, it is advisable to let them have a little light dry grain. This is especially necessary in winter: the warmth it gives prepares them for the long winter night, and if accustomed to this meal, and it should by any chance be forgotten, they will wait in the accustomed place long after roosting hours, as if confident that their empty crops would not be forgotten.

Such is the bill of fare which we have found to suit, and we think that, were it adopted more generally, we should seldomer hear the complaint, " Poultry do not pay." Nevertheless, we would be far from limiting ourselves to it, or advising others to do so. Variety is one great charm of life; variety we demand in our own tables, and variety we must give also to our feathered favourites, if we would have them fulfil our expectations. Our own appetites would flag were the same dish to be presented to us day after day, excellent though that dish might be; in like manner we will find our poultry droop, if we present to them the same unvarying mess. Let the hen-wife, then, study a reasonable variety in their food. And she has a wide range for this—oats, barley, wheat, rice, may all be given in turn. Buckwheat is much used in France for fattening fowls. It is grown successfully in England, but not in Scotland.

Sunflower seeds are said to fatten fowls rapidly, and are eaten greedily by them. In America and Lombardy, and in the southern parts of Germany, Indian corn is employed for feeding poultry, and they become very fat upon it. It requires to be boiled, as the grain is too large to be swallowed whole, and will be found a cheap as well as nourishing food. Rice may be given raw; the fowls will pick it up like any other grain; but it will be found a much more economical plan to boil it, for when boiled very soft and allowed to cool in the water, it will swell out to an amazing extent. It is more expensive than grain, except when it is got damaged from wholesale dealers, in which case it will be found economical. Grain can be given raw and boiled alternately, except oats, which must always be given in its raw dry state. Barley, if given raw, must be supplied very sparingly, for it swells out so much in the crops that, if given freely, it will be sure to injure the fowls. The safest plan is to have it boiled, so that the process of swelling may be accomplished before it enters the crop. In this way it will be found an economical food. Wheat is also rendered more economical by boiling. It is a favourite with fowls. Whatever be the method adopted of feeding fowls, they should have a fair allowance of dry grain each day—either a good feed in the middle of the day, or a smaller quantity twice a day, morning and evening. If they have too much soft and green food they will be apt to have diarrhœa; if, on the contrary, they have too much dry grain, they will suffer in the opposite way. Attention must be paid to the proper regulation of the diet. By watchfulness in this respect disease and death might frequently be warded off.

Fowls are very fond of bread, as may be seen by the eagerness with which they scramble for the crumbs occasionally thrown to them; the scraps, therefore, should be carefully preserved for them, and the hard crusts should be steeped and mixed with their soft mess. Any rough bone may also be thrown to them; they will pick it very neatly. They display carnivorous propensities to a considerable extent — so much so that they have been seen watching for a mouse at the taking down of a stack, and, the moment it tried to escape, running after it in full chase— cocks and hens together—and not only pecking it to death, but actually devouring it.

The bill of fare we have presented has been intended for the poultry-yard in general; but should there be any young hens of the early broods, which in autumn show a disposition to lay, and should the hen-wife wish to encourage them in this, she can indulge them with more generous fare, and more comfortable housing. The feeding may be warm potatoes or Indian corn, with firm oatmeal porridge twice a-day, morning and evening, with a little dry oats at mid-day. The housing may be in the hatching-house, where warm nests of clean oat-straw may be prepared for them, furnished with nest-eggs. They must be separated from the others at feeding-time, so that they may have an opportunity of enjoying their tempting fare in peace. The increased redness of the comb and the singing noise which the hen makes, indicate that she will soon begin to lay. As has been previously stated, the laying-hen must have access to limy rubbish, in order to form the outer covering of the egg. Like every workman, she must have materials; if we do not supply them to her, we cannot expect any produce. Fowls of all kinds require sand or gravel to assist in grinding the food in their stomach. We have read of certain fowls on board ship, which, notwithstanding all the care bestowed upon them, sickened and died in their coops. On being dissected, the want of any gravelly substance was found to be the cause of their death. Stones being obtained at a convenient port, were ground down and given to the remaining fowls, which, although affected in the same manner as those which had died, soon returned to a healthy condition. The dust-bath, too, must be kept in mind, for fowls suffer very much if deprived of this opportunity of freeing themselves from those troublesome parasites which are apt to infest them, and which, if allowed free quarters, form the bane of their existence. And water—pure fresh water—must not be forgotten; it must be daily supplied, and supplied, too, in abundant quantity. The vessel which contains it must be daily cleansed. The feeding-trays must also be kept clean; if this is not attended to, the food will become sour, and the poultry will not thrive. Scrupulous cleanliness is indispensable in the management of poultry. The hen-house and yards must be kept free from impurity. Some adopt the method of scattering daily a little straw over the floor of the hen-house, sufficient to cover the droppings, and to clean out the whole once a-week. But we think that a daily sweeping of the floor, followed up by a liberal sprinkling of sand and gravel, will be found an *easier*, as it certainly is a more cleanly plan. The yard, if of gravel, should be frequently swept; if of earth, it should be dug occasionally. The worms exhumed by this process are highly relished by the fowls, and the opportunity it gives of scraping in the soft fresh earth is very advantageous. The hen-house should be well aired every day; the fowls will usually resort to it of their own accord at roosting-time; but the hen-wife will do well to see that all are safe, and no straggler absent, before she shuts them up at night. This she must do regularly to protect them from the inroads of their natural enemies of the quadruped kind, as well as from those predatory attacks of a different nature, which the hen-wife has but too much cause to fear. In winter, every inlet to the hen-house must be closed at night, but, in summer, it is often convenient to leave a small outlet for the hens—they are early risers, and often chose to leave their roosts before the hen-wife is astir. The danger from rats and similar vermin is not great at that season. If kept shut up in the morning beyond their usual hour, the commotion they make is something surprising; and in the scuffle that is sure to ensue, they are apt to injure each other.

If poultry be treated in the manner recommended above, we are sure that in a yard of any size there will at all times be found fowls fit for the table, in good condition, at any hour they may be required, and without any formal preparation. Fowls fed in this manner, which we may call the natural method, will generally be preferred to those which are subjected to the cruel and unnatural process known as the " cramming " system. Nevertheless, for the benefit of those who wish to adopt it, we say a few words on the subject. To place a fowl in total darkness—to tie up its legs so that it cannot possibly move—to put its eyes out even, so that it may be kept in more perfect repose—is a plan so cruel that we could recommend it to no one, nor do we think any person with ordinary feelings of humanity would adopt it, even were we to do so. Every right-minded person must shrink from such cruelty. Though allowed to use the animal creation, we are not permitted to torment them; and assuredly we ought to make their little life as happy, and their death as easy, as circumstances permit.

Without adopting the cruel parts of the process, however, it may be used in a modified form, and thus be found useful in improving the condition of fowls. Without having its eyes put out or its feet tied, the fowl may be penned up in a small coop. It may be stuffed once a-day with a paste composed of barley-meal, coarse sugar, suet, and milk. It must be allowed to eat as much as it likes, in the first place, and then the paste must be *forced* down its throat till it can literally contain no more. It must then be left in *the dark* till next feeding-time. At the end of a fortnight it will be found to have increased enormously, and must then be used. On no account must it be kept longer in the feeding-pen than the above-mentioned time. The state of repletion in which it is kept is apt to induce fever if continued long, and thus to render it unfit for the table.

Where speedy fattening is required, the following plan, extracted from the "Poultry Kalendar," will be found excellent :—

"Put the fowls to be fatted into a quiet place, and feed them three times a-day with rice, boiled in skim milk till it is quite soft and swelled out. Give each time just as much as will satisfy them, and as dry as possible. Remove the feeding-dish each time. Let it be thoroughly washed, that no sourness may be conveyed to the fowls, as that prevents them fattening. Give them the milk of the rice to drink, or a little clean water. It will be found that this method gives the flesh a whiteness which no other food can give; and when it is considered how short a time is required for fattening, it will be found as economical as any other mode, or even more so. In five or six days the fowls will be sufficiently fat for the table. The coop in which they are confined must be cleaned daily, and no food given for sixteen hours previously to being killed. Of course the fowls must have gained size and flesh previous to being put up, or the process described above will be of no use."

Another excellent plan we extract from the same work:—

" Mix barley-meal with water or skim milk into the consistency of cream; put a little coarse sugar or treacle into the mixture; put it in a shallow dish beside the fowls; do not give them any water—thirst induces them to consume large quantities of the mixture; let them have as much as they want; change or clean the dish three times a-day, to prevent the food getting sour, as sourness prevents fattening. They must have sufficient light to see their food, but the less they can do with the better; put a little sand or gravel in the coop."

The last method we propose is also from the " Poultry Kalendar ":—

" Give oatmeal and barley-meal alternately, mixed with milk and a little dripping; let the dish be cleaned, as in the previous directions; gravel and sand supplied, and, in this case, abundance of clean water given."

It may here be stated that, in order to insure the success of the above, or of any other methods of fattening, it is necessary that the fowl should previously have acquired both size and flesh—feeding first, and fattening afterwards—a time for each, and each in its time, if we would have success.

CHAPTER V.

ON THE HATCHING AND REARING OF CHICKENS.

WE have now arrived at the most important stage in our poultry career—a stage in which all the energy and activity of the hen-wife must be brought into play. Hitherto the work may have been done, and done respectably even, by a hen-wife in no ways famed for these qualities; nay, her inactivity may even border on laziness, and yet, with adult occupants of the poultry-yard, her success may be unimpeachable. But with the rising generation of the feathered tribe, the case is altogether different; here activity is indispensable—activity kept in continual operation by an interest that never flags. The hen-wife who grudges to leave her sofa or her easy-chair at the call of duty, had better consider ere she commences an undertaking which, in all probability, she would soon relinquish in disgust; and she, who scorns the idea of superintending in person the operations of her poultry-yard as derogatory to her dignity, would do well to follow the same plan.

Were such cases universal we might lay down our pen in despair, but knowing, as we do, that many, high in station as in intellect, have not thought it inconsistent with their position to engage in the pursuit of which we are now treating, and believing that there are many such from that station downwards, we continue our labours, in the hope that they may be as profitable to others as they are pleasant to ourselves. And let it not be supposed that we mean to intimate that the work of this busy season is either troublesome or difficult. The labour it entails is so linked with delight as to exclude the idea of trouble, and the success which is almost sure to attend patient and rational care banishes that of difficulty. The hen-wife, then, need have no fear on this score, but may commence the interesting labours of this stirring season, in the hope of soon seeing herself surrounded by numerous and thriving broods. She will watch with interest for the clucking sound that indicates the desire to brood, and, by providing comfortable nests furnished with tempting eggs, she will give every encouragement to the hen to sit. She may reckon herself fortunate if she have any inclined to do so in the beginning of the year, and will lose no time in setting them, whatever be their qualifications as mothers, as it will give her an opportunity of rearing early broods, the cockerels of which will come in early for table, and the pullets will serve as layers in the ensuing autumn and winter. In March and April there will be numerous chickens, and the hen-wife can then afford to reject those which are unsuitable. An ill-grown, ill-feathered, ill-tempered, and ill-conditioned hen, must not be allowed the privileges of maternity, however much she may desire to have them. In all probability she would only bring out a few sickly chicks, and even these she might fail to rear; her time, therefore, would be all but lost, whereas, notwithstanding her defects, she might produce eggs, deficient neither in quantity nor quality. To reserve such for layers would therefore be more profitable; and, passing them by, the hen-wife may choose, as best suited for her purpose, a fat, well-feathered hen; such will be best able to cover and keep warm the brood. She must be of a quiet, gentle disposition, not easily frightened, and not disposed to wander; for if she remain off her eggs too long they will become cold, and thus the principle of life will be quenched. If she have previously proved

herself a good mother—neither careless nor too solici-
tous about her brood—so much the better. Careful
she must be, but not to excess, else even the necessary presence of the hen-wife will give her alarm,
and make her fancy that harm is intended to her
progeny; and thus, while defending them from fancied
danger, she will probably inflict upon them real and
perhaps fatal injury. It has been recommended by
some to set only one hen at a time, in order that
there may be a succession of fowls for the table;
others have preferred setting two at once, so that if
one or both broods should prove deficient, they might
be given to one hen, and the other returned to the
yard. Both of these plans are on too limited a scale
to suit our purpose, which is to produce an abundant
supply. Even in ordinary seasons the market would
receive what the family table did not require; much
more so at this time will the scarcity with which we
are threatened make but too ready a sale for all, and
more than all, that we can produce.

The hen-wife, then, will do well to encourage every
hen that has the necessary qualifications for maternity; and, in order that she may do so, will not delay
to provide her with eggs. Large, well-shaped, fresh
eggs should be chosen, not too thick in the shell, nor

yet too thin, and with the substance uniformly diffused throughout the egg. Double-yoked eggs should be rejected: were they successfully hatched, two chickens in each would be the result; but as we suspect this rarely if ever happens, it is better not to risk the loss of the egg, and the disappointment that would ensue. The eggs selected must next be examined. This is best done by the light of a candle. If the air-vessel at the broad end of the egg is clearly seen, the egg may be set, if not, it is unproductive. About this, however, we must say there is a diversity of opinion; some asserting that the presence of the air-vessel is no proof of the fertility of the egg. It has been affirmed by some, that if the air-bag be at the extremity of the egg, it will produce a male bird; and if at the side, a female. For the truth of this theory, however, we cannot vouch, not having tested it. If the eggs of any particular hen are wanted for setting, they must be kept by themselves until the hen has produced a sufficient quantity. They may possibly continue productive for two or three weeks; but, of course, the fresher they are, the greater is the likelihood of their being hatched. They should be kept in a warm place, with the pointed end undermost.

E

Our cluckers being now chosen, and eggs provided for them, the next part of our process is to prepare the nests. Now, we all know that when a hen sets herself, it is always at the root of a tree or hedge, or in some corner where the foundation is the damp ground. We all know, too, what healthy birds she brings in triumph to the homestead, claiming, as it were, their right to a maintenance. Now, if we would have Nature's success, we must imitate as far as we can her procedure. We would, then, take a fresh turf as the foundation of the nest. The advantage of it is that it prevents the feverish heat which a close dry nest induces, and which often weakens the chickens so much as to render them unable to leave the shell. If turf cannot be had, damp sandy earth will supply the defect: we have tried it with success. Above this we would put a plentiful supply of good oat-straw, cut into small pieces. We use straw in preference to hay, as being less liable to get musty; we cut it to prevent the long straws getting entangled in the feet and necks of the chickens, which they are apt to do, and so to hurt and even kill them. This nest we would place literally on the ground, and put it into shape by means of a few bricks, stones, pieces of wood, or anything that comes to hand. Any out-house that could be spared (dry and free from draughts, if possible) would do for this primitive nest; and, with a division between each, a dozen might be accommodated in a moderately-sized place. If more space were wanted, any spare cellar or loft, or even the corner of a washing-house or scullery, would do. We have even seen an old barrel turned on its side, well lined with straw, and covered with tarpaulin, used for this purpose. Surely it would be difficult to find the purse or the premises which this plan would not suit.

To those who are inclined to be more fastidious in their choice, we would recommend, as the best of all, wicker-work baskets, of a round form, widening gradually upwards. The bottom of the basket must just be large enough to admit the body of the hen; its widening upwards serves to admit her wings and tail. For this last there must be sufficient space, else it will be twisted, and, if so, it will render the hen uncomfortable, and thus impede her work. If square boxes be used, the corners must be filled up, to prevent the eggs and chickens rolling into them. The nests, of whatever kind, must be placed apart from the other fowls, as the hens will not sit unless they have quietness and repose.

The number of eggs under each hen depends on the size of the eggs, the size of the hen, and the season of the year—*eleven* and *thirteen* are the numbers which custom sanctions. A large hen might cover either of these numbers, and hatch them too, provided the weather were mild; still, with fewer, even in favourable circumstances, we would think the chances greater. In the beginning of the year, when frost and snow may be expected, a much smaller number will be sufficient—seven at the most; for provision must be made, not only for the hatching of the eggs, but for the brooding of the chickens. Now, as early-hatched chickens attain a considerable size before the setting-in of genial weather, the hen must have no more than she can cover; if she has, the outside chicks get chilled, they squeeze themselves in, forcing others out, which in their turn get chilled, till the whole in succession suffer, and one by one they will either droop and die, or grow up stunted and dwarfish things. It were better, then, to have fewer, and rear them all. Warmth, above all things, is necessary for chickens.

The sitting-hen must now be attended to. When she begins to cluck, she usually makes choice of a nest—one, perhaps, with an unusual number of eggs in it; these taken from her, she still sits with the same assiduity, and it is with great difficulty that she can be persuaded to sit in that destined for her. The best plan is to try her first with addled eggs; place her gently on them, and leave her in the dark; of course night is the most suitable time to do this, natural being preferable to artificial darkness. The following night, the nest being thoroughly warmed, the eggs selected for the purpose may be given her. It is well to remove the hen to the hatching-house at an early stage of broodiness, otherwise the difficulty of removing her may be found insuperable. Having fairly taken to her nest, she requires little of either food or attention during the time she so patiently performs her apparently wearisome task; that little, then, the hen-wife will bestow with kindness and regularity. She will visit her at the same hour every day (the morning is the best time), supplying her with food and clean water, and seeing that the dust-bath is within her reach. The food we have found most suitable to the sitting-hen is dry oats. Soft food, such as we have previously described, has been re-commended; to this there can be no objection, provided dry grain be also added. The hen will, in all probability, voluntarily leave her nest and return also of her own accord. Should it be otherwise, the hen-wife must lift her gently off, and direct her to her food. Should there be a small yard contiguous to the sitting-house, a run in it will be of advantage to the hen, but it is not advisable to let her at large, as she might wander too far, and allow her eggs to cool. The same individual should attend the hen, as she would be sure to be frightened by strangers, and to break her eggs by the sudden movements caused by terror. Children must not be allowed to rush out and in where they are sitting, or to make loud and sudden noises in their vicinity. If they do, you will look in vain for your chickens.

At the end of the twenty-first day there will be a commotion in the nest; and if all has gone well, the heads of several little chickens will be found here and there peeping from under the wings of the mother-hen. This is a pleasant sight to the hen-wife, who now begins to reap the reward of her cares. She will gently remove the egg-shells and make the nest as comfortable as possible. Sometimes it is necessary to chip one side of the egg, and assist the chick to leave its prison. Dorkings especially often require this, as the shell is so thick; but, as a general rule, the less they are interfered with the better.

The hen should be fed, and the chickens left undisturbed for nearly twenty-four hours. At the end of that time they may be fed with dry oatmeal and water. With one hand the hen-wife can open the mouth of the chick, and with the other put into it some oatmeal,—being dry, there will be no danger of choking; the bill can then be dipped into a saucer of clean water. This was our first plan. An observer was highly amused by it, saying it was vastly more like " forcing up " than " bringing up." It " brought up " admirably, however; still there is no necessity for the hen-wife taking this extra trouble unless she is inclined to do so, for by waiting a little longer the chicks would learn to pick for themselves. But in this case the oatmeal must be moistened with a little milk or water, not too wet, but of a *friable* consistency. And to prevent the chicks wetting or drowning themselves in the water, it may be put into a saucer which has been half-filled with small stones; or, what is still better, a vessel of earthenware used for forcing early mustard may be employed. It is in the form of a saucer, with concentric troughs of equal width; these troughs receive a sufficient quantity of water, and, being only about an inch wide, there is no danger of the chicks either wetting or drowning themselves. This we have found to be the best possible drinking-vessel for chickens and turkey poults, and heartily recommend it to every hen-wife. The hen meanwhile must be supplied with corn; she may also have from time to time a portion of the soft food previously described. The scale which is on the extremity of the chicken's bill may be removed with the finger nail, so as to enable it the better to pick up its food. The nest must be taken out of the box or basket, and placed on the ground, with a slope of straw from it, so that the chickens may walk up to be brooded. Should the nest be originally on the ground, the bricks and stones, and everything that might cause danger to the chicks, must be removed. They must be visited very frequently and supplied with a variety of fresh food,—crumbs of bread, boiled rice, picks of porridge, and potatoes—all will be welcome, and, being fresh and clean, will be heartily devoured. Clean water must never be forgotten. Of buttermilk they are very fond, but it must be given in addition to water and not instead of it. Their last meal should be at nightfall; the eager haste with which they rush from under the wing at the approach of the hen-wife is really amusing, and the appetite with which they consume their evening meal shows that it is not unnecessary. They are also early astir in the morning, and if the hen-wife is not a very early riser, she should leave them something to pick at, as it is not unusual for them to be looking out for something at four in the morning. In a week tailings of wheat may be given, or even ordinary grain. The mother-hen will break it down for them, and even feed them. From the earliest days of chickenhood green food must be supplied—a cabbage, or lettuce, or turnip top, or even grass minced down and mixed with their food; they will pick it out and eat it with apparent relish. But when the weather is favourable, and they are allowed to walk abroad, as they may do under certain restrictions when they are a week old, they will enjoy such food with more zest, picking it for themselves in the green spots to which they have access.

If at this stage the hen be placed under a coop on a grass run, it will be found very advantageous. Should that not be practicable, she might with safety be admitted into the garden. The little injury chickens do is amply compensated by the good they accomplish in picking up the worms, grubs, and insects. The hen, being confined in the coop, would not have the opportunity of making her dust-bath

among the flowers, which otherwise she would be sure
to do, making sad havoc among them. Food and
water must be placed outside the coop, within reach
of the hen, and the chickens, however far they may
wander, will return at her well-known call. Should
there be numerous broods, the hens will be apt to fight,
and thus injure the poor little things that flock around
them. Their love is so exclusive that they cannot
bear to see the progeny of another have any enjoyment, and so will be sure to peck at the strange chicks,
and prevent them feeding if they can. The hens
should therefore be confined in separate coops, but
in close vicinity. When they get accustomed to each
other, they may, if there is convenience for it, walk
forth at liberty, each surrounded by her own confiding family, secure that, however far they may wander,
her call, distinguished from every other, will bring
them without fail under her sheltering wing. The
coops, it should be stated, must be moved to a fresh
spot each day. Should they have wooden floors, they
must be cleaned daily, and well sanded or gravelled.
Dorking chickens, however, must not be reared on
wooden floors. They do not thrive on them; disease
of the feet is induced thereby. Chickens, and hens
too, get attached to their dwelling-place, so that the

less they are moved about the better. It ought to have been stated that, where any of the sitters fail to bring forth families of a respectable size, two or more broods may be put to one hen. The bereaved mothers must be shut up in solitary confinement till they forget their lost ones, which will be in the course of a few days.

By the above mode of treatment, continued for a fortnight, or until the wing and tail-feathers begin to grow, the chickens will be so far advanced as to admit the hen-wife to remit many of her attentions. Instead of visiting and feeding them every three hours, three times a-day will now be sufficient, and the same food may be given them as that supplied to the larger fowls. When the hen has continued tending her brood six or eight weeks, she will begin to lay again, and will probably bring out a second brood towards autumn.

Although March is the principal month for setting hens, it is by no means the only one. In April and May the peculiar sound of the brooding-hen will be heard from time to time, and the hen-wife who would have a succession of table-fowls will avail herself of the opportunity thus afforded of hatching them.

But in June the thoughtful hen-wife, whatever be

the opportunity she has, will act differently. It has been observed that chickens set at that time have uniformly been attacked with a fatal disease at their first moulting season. Among other symptoms, the skin has been observed to be red and inflamed, caused, it has been ascertained, by hundreds of harvest-bugs having effected a lodgment there. The irritation and suffering they cause is very great, and very few of the poor little things recover. The only remedy discovered is to anoint them with oil and vinegar; but even this is rarely successful. The best plan is to avoid hatching at that season. Should there be any particular reason for doing so—any rare and choice eggs, for example—and the hen-wife resolve to risk it, she must choose a *very cool* situation for the hen. She might also watch with care the first symptoms of the disease in the chicks, and lose no time in applying the above-mentioned remedy.

In July and August there will again be numerous cluckers, and the hen-wife will have an opportunity of filling up, in measure, the vacancies in her yard. The same method of procedure as has been recommended above may again be adopted. Greater attention, if possible, must be paid to the chicks, that they may gain size and strength before the frosty mornings and evenings begin. It is generally found to be more difficult to bring autumn chicks into good condition than those hatched in spring; but when they *are* up to the mark, they come in seasonably for Christmas festivities—at once convenient for the family table, and bringing a high price in the market. As chickens of all kinds are injured by rain and heavy dew, they must not be let out early in the morning, and when heavy clouds portend a shower they should be brought under cover. If overtaken by a sudden storm, the mother-hen shows the most affectionate solicitude, extending her wings over her brood, bearing patiently the pitiless pelting of the storm, content if they are safe. What a touching spectacle, and what an instructive lesson! What a vivid commentary, too, on the sacred text,—"O Jerusalem, Jerusalem, how often would I have gathered thy children together, *even* as a hen gathereth her chickens under her wings, and ye would not!" Could an illustration more expressive of intense love have been chosen? Yet Israel heeded not, and is now scattered among all nations, a reproach and a by-word, trodden down, despised, and oppressed.

CHAPTER VI.

MISCELLANEOUS HINTS.

BESIDES carrying out her arrangements so as to have an abundant supply of fowls for table and for market, the hen-wife must also arrange so as to have an abundant supply of eggs also. This is an article of food so suited to the child and to the invalid, and one from which such a variety of nourishing and palatable dishes can be made, that it must on no account be overlooked. It is comparatively easy to secure a supply during summer; but in winter the case is very different. It is quite possible, by more generous food and warmer housing, to induce the hens to go on laying a little longer than they would otherwise do, but this of itself will not make them do so all the year round. Indeed, if overdone, it will have quite a contrary effect, for nothing more effectually prevents a hen laying than over-feeding. A fat hen, as everybody knows, will not lay at all.

Evidently, then, this is not the plan, and some other must be devised; fortunately, we are able to propose one which cannot fail of success—and this is simply to take advantage of the natural laws to which fowls are subject. Hens begin to lay at a certain age, some sooner, some later, according to the kind. Cochins begin about five months old; Dorkings and Spanish about seven months. Now, by having broods so arranged that their pullets will be ready to lay when the older ones have ceased, a supply for the winter will be secured. It will be advisable, as the successive broods advance, to select from them the pullets which are likely to become good laying hens, and reserve them for that purpose. Plump, full-breasted, lively birds must be chosen, with small heads and moderately long grey legs. The colour is a matter of taste; those of a dark colour are generally supposed to be the best layers. When these pullets begin to lay, as they will do in succession, according to their respective ages, they must be treated according to the plan laid down in a former part of this work (Chapter IV., p. 62). They will lay all winter; and in order that they may do so during the summer also, the hen-wife must continue to feed them well and regularly; for if they are obliged to

go into the fields to seek their food, they will be apt
to make nests for themselves, lay, and hatch broods,
which would defeat the hen-wife's object.   In winter,
being no longer pullets, they will become subject to
the laws which govern hens: they will cease to lay
in winter, and may be used the following season for
mothers, should they possess the qualifications for
that important position; while the spring-hatched
pullets will be, in their turn, ready to fill the place
occupied by these hens the previous winter.   A constant supply, year after year, will thus be kept up.
The broods thus thinned by the abstraction of the
pullets must be still further reduced as they grow up
towards maturity.   The cockerel, as soon as ready,
must be despatched to the cook or to the market.
As has been already said, every grain it eats after
this is just so much thrown to the winds.   The pullets not intended for layers should also share the
same fate.   At three months old, the sexes should be
separated.   Should any of the cockerels be suited for
roosters, they should be reserved.   The hen-wife
should at all times be careful to gather in the eggs as
soon as laid; this should especially be attended to in
summer, as they are apt to be addled by the hens
sitting on the nests.   They should also be sent off at

once to market, as they change rapidly during warm and moist weather. Care must be taken in rainy weather while bringing the eggs from the hen-house; a single drop of water is sufficient to taint the egg, if suffered to remain. Keep them dry, therefore, if possible; if not, wipe them gently with a soft towel. In the season when eggs are most plentiful, it is a good plan to preserve some for winter use. There are several ways of doing this. Anything that excludes the air answers the purpose, such as bran or salt; the pointed ends must be placed undermost, and they should be closely packed. Lime-water is also much recommended. The lime is dissolved in water, and, after standing some days, the clear portion is poured over the eggs which have been previously packed, with the pointed end undermost, in an earthen jar. Eggs preserved in any of the above-mentioned ways, do quite well for puddings in winter, and those that are new-laid can be reserved for using in the shell. But the method which we think the best, and at the same time the simplest, is to smear the shell slightly with good fresh butter, and place the eggs, as above, in a jar. They will keep for several months, and at the end of that time will retain the delicious milkiness peculiar to a new-laid egg. It ought to be done as soon as possible after the egg is laid. It is rather surprising that a plan so efficacious for procuring a fresh-laid egg is not more generally adopted. The egg, it need scarcely be added, is unproductive after being smeared or put in lime.

Having spoken of overfeeding as being so injurious, the hen-wife may probably expect that we should give some idea of the quantity requisite for a given number of fowls. We do not wonder if she does, for we well remember our own anxiety on this point when we commenced our poultry career. We well remember how eagerly we asked, "How much must we give?" and the answer we also remember (given by a very practical person), "You will very soon find out that." And truly we very soon *did* find it; and so will the hen-wife who faithfully tends her confiding flock. If she errs on the side of prodigality, the unused food will teach her to give in future more stinted supplies. Should she err on the opposite side, the mistake will be even more readily discovered. The empty crops and eager looks of her feathered dependants will convey to her a sense of their wants in language which cannot be misunderstood.

And with regard to calculations of the cost of food

for a given number of fowls, though curious and in-teresting in their way, we have never experienced any benefit from them, nor do we think others will do so. It cannot guide the hen-wife half so well as practice and experience will do. It may show her, indeed, at how trifling a cost a chicken may be reared, and the immense profit that must accrue to the hen-wife who rears on a large scale; but if she follows the method we have been advocating, she will have no need to learn this from extraneous sources. Her own rapidly-increasing profits will teach her the les-son more effectually, and at the same time more plea-santly.

Fowls of all kinds being natives of warm climates, warmth is a condition that cannot be dispensed with, if we wish their prosperity. This they find in the genial warmth of spring, when they will be seen sunning themselves at noon-day. In the height of summer, again, they seek the shade, for extremes of either heat or cold are injurious to them. As the season advances, and the cold increases, the hen-wife must see that they have the benefit of sunshine when we are favoured with it. She must also make up in some measure for the want of summer heat, by ad-ministering more generous food—perhaps an extra allowance of grain, or a more liberal supply of animal food.

During the moulting season especially, which is in autumn, they must be more liberally fed, and their housing then must be well attended to. Cold draughts must be carefully excluded, and they must be protected from rain during the day as well as at night. It cannot be too well kept in mind that a dry, well-drained, well gravelled yard is of the greatest possible consequence to them. In a damp, wet yard they cannot possibly thrive; it is miserable to see them in such a puddle.

As the season advances the hen-wife will continue still further to thin her coops. She will find it more difficult to fill the crops of the pullets and cockerels than when they were chickens, and so will hasten to dispose of them as soon as ready. The cockerels, of course, must suffer first, for there is a natural repugnance to kill pullets which might turn out good layers. Instead of doing this, the thrifty hen-wife will rather take the old hens for her own table. By keeping them for a week after being killed, they will become sufficiently tender for boiling. If very old and incurably tough, they will at least assist in the preparation of that soup so well known in Scotland by the name of "Cockie-leekie." But while old fowls are decidedly improved by being kept some days after being killed, a young fat fowl is never more delicate and tender than when killed immediately before being dressed for table. A young male bird makes the best roast, and the young female the neatest boiled dish. A hen, to be good for table, must be fat; she must have a plump breast, fat under the wings and under the skin, and her flesh must be white and her fat firm. The scales on the legs of a young hen are smooth and glossy; she has only the rudiments of spurs, and her under-bill is soft. The scales on the legs of an old hen, on the contrary, are rough, and her spurs are hard; her under-bill, too, is quite stiff, and her comb thick and ₁ough. A Cochin should not be used for table after it is four or five months old; it becomes coarse after that time. The Cochin pullet makes the best dish, as the cockerel has too little breast meat. Fowls of other varieties are tender for a much longer period. Verging on a year old, their tenderness is doubtful; and after it they are considered *old*. The best plan is to keep them always young. It will be found the *great secret* of making them profitable.

During the time of a snow storm, it will be necessary to keep the fowls in the hen-house, or in a covered shed. It is curious to observe the stupefying effect which snow has upon them. But should it continue to remain any length of time on the ground, they will gradually get accustomed to it, and may be let out, as long confinement would be injurious.

It will be necessary to state that the cocks from which the breed is desired must be chosen several months before hatching time, and all others removed. The exact time cannot be stated, but the safest plan is to have at all times the best roosters in the hens' yard, and of the kind wanted. It has been already stated that a cock of a totally different strain must be procured every second year, for breeding in-and-in, as it is called, must by all means be avoided; it will cause the very best kind to degenerate, and will effectually prevent poultry from being profitable.

The plan to be adopted with a clucking-hen, who persists in sitting when she does not possess the necessary qualifications, may here be mentioned. The cruel and dangerous custom of sousing in cold water we unhesitatingly condemn. Shutting her up in a dark cellar without a nest for two days without food, or with a scanty supply, will be found quite severe enough, and will generally be found effectual. It has been recommended to put her under a light tub, made for the purpose, pierced with holes to admit the air, these holes having such an inclination as to exclude bright light. The tub should be of a size just sufficient to admit one hen standing. This is certainly more merciful than whelming a large tub over two or three hens, where they have light enough to peck each other to death.

The disposal and management of the feathers is another thing that calls for the attention of the henwife. As soon as a fowl is killed, and while yet warm, let it be carefully plucked. Separate the large wing feathers; put the others into small paper-bags previously prepared. Put these bags into an oven, and let them remain about half-an-hour; take them out, repeat the process two or three times, then keep the feathers in a dry place till required. The oven must not be too hot. Care must be taken to free the feathers of any skin or flesh that may adhere to them while being plucked, or they will be tainted. The hard quilly portion of the larger feathers must be cut off with a pair of scissors. The wing and tail feathers may be stripped and added to the others. Previous to putting them in the oven, some recommend that the feathers should be put loosely into a dry tub or basket, and shaken up daily, so that all may in turn be exposed to the air. Others recommend, as an easier plan, merely to suspend the bags from the roof of a warm kitchen, or on the wall behind a fire-place, where it is practicable. In this case they will take longer to dry. Feathers can be quickly and effectually dried and cleaned by the agency of steam; but it is rather an expensive method, and the thrifty hen-wife will doubtless prefer having the produce of her own yard prepared under her own eye, and by her own directions.

The fowls whose good fortune it is to belong to a farm-steading, are usually turned out into the stubble when the crops are removed. With such sumptuous feeding, they will thrive amazingly and contribute greatly to the contents of the egg-basket. Those which are not so favourably situated will be found improved by being turned into the kitchen-garden for a few hours each day. When the potatoes are taken up, and much of the green crop consumed, they will not do much harm; and the little injury they do will be amply compensated by the additional fertility given by their droppings to the soil. If in a confined situation, the fresh air sets them up wonderfully, and prepares them for the winter; while the worms and insects they find, contribute without any extra expense to their egg-producing powers.

Hitherto we have made no mention of the diseases to which fowls are subject. The hen-wife who carefully follows the plan we have laid down, and who provides for her poultry warm housing and proper food,—who attends to cleanliness, air, and rational treatment,—will have no cause, we are persuaded, to complain of disease in her yard. And should it appear, the remedies which have been propounded would not, we fear, aid her in its removal. If disease cannot be prevented, we have little hope of its being cured. The simple remedy of mixing a little Epsom salts or jalap in their food, on the first appearance of illness—such as hanging the wing, delaying to leave the hen-house, etc.,—might do good, and could not possibly do any harm. Some give this occasionally to all their poultry, and find that it does them good. It should be in the proportion of a tea-spoonful to twenty fowls.

With chickens we have seldom found nursing do any good; occasionally we have seen them restored, when apparently dead, by the warmth of a fire, or even by that of the hand. But the best plan is, by unwearied attention, to ward off disease, for prevention is at all times better than cure.

CHAPTER VII.

TURKEYS—GEESE—DUCKS, ETC.

A WORK such as this would not be complete did we confine our attention to the domestic hen. There are several other kinds of domesticated fowls which deserve notice, and which the hen-wife would do well largely to cultivate and rear. Among these the turkey first deserves attention. Originally from the wilds of North America, it is now completely naturalised in this country, and forms a most valuable addition to our poultry-yard. The cock is a stately and majestic bird, and when he struts about, trailing his wings, he has a noble and martial bearing. But when he is roused, he is the very perfection of passion; at least he gets the credit of it, and certainly his appearance betokens it. The hen, on the contrary, is distinguished by her gentleness, except when her poults are interfered with, and then, indeed, she might vie in fierceness with the she-bear bereaved of her whelps. Her appearance is very elegant, and she is, as one of our friends remarked, " a very lady-like bird," and, on the whole, forms a beautiful addition to our feathered stock. But the beauty of the turkey is not its only advantage. It is justly prized as a table bird, and its flesh is esteemed a great delicacy.

At the festive season it is in great demand; it occupies a conspicuous place at every friendly gathering; indeed the table would not be considered complete without its portly presence. The great objection is the difficulty which is said to exist in rearing the young, for, when reared, the turkey is allowed to be the most hardy of animals. This, although almost the universal opinion, is, we are persuaded, a most erroneous one. A high authority gives it as his opinion that the young of turkeys are as easily reared as those of other fowls. His experience of twenty years confirms his opinion, for during that long period, with two hatchings each year, not one death occurred. The treatment that was so successful he describes—to which in its proper place we shall refer. This favourable opinion also receives further confirmation from what is seen every day in Ireland. There, in the vicinity of every cottage, you will see them in scores—nay, in hundreds—picking up a living in the fields, on the hills, on the commons, on the roadsides even. The small farmers rear them in large numbers; they share the children's breakfast of oatmeal or Indian corn; and so profitable are they, that the landlord's rent is almost paid by the sums that they realise. Surely, in such circumstances, the attention they receive cannot be very great, and proves that they can be reared with comparative ease. The greater mildness of the climate is certainly one advantage, and the dry earthen floors are also in their favour. Nevertheless, even in this less favoured climate, we are persuaded that, with moderate care and rational treatment, they might be successfully reared, and in such numbers too as might give the hen-wife a handsome return for her outlay. To assist her in doing so is the object of the following remarks.

The turkey-hen begins to lay in spring, generally in the month of March. She will lay from twelve to fifteen eggs; and, like other fowls, when she has done laying, will be disposed to sit. In Ireland it is the custom to withdraw the eggs after half-a-dozen have been laid, to induce the hen to go on laying a greater number. This plan is often successful to an extent we could scarcely credit. An instance is on record where a turkey-hen went on till she had laid the enormous number of ninety eggs. This plan is not confined, however, to turkeys, for in the African deserts the natives adopt the same method with regard to the ostrich.

The egg of the turkey is considered a great delicacy, but in Britain they are considered too valuable to be eaten. In Ireland, however, they are to be found in tolerable abundance in the market—so much so, at least, as to be sold at the low price of 9d. per dozen.

The nature of the turkey inclining her to wander, she will be apt to make a nest for herself among the trees, shrubs, or plants which may be in her vicinity. Having once laid in the place of her choice, she will continue to do so, even should it be discovered, and the egg removed from time to time. A high fence or wall intervening will be no obstacle in her way, for she flies to a great height. Care should, then, be taken, just before the laying season commences, to provide her with a comfortable nest, of proportions suited to her size. It must be in a quiet retired place, for the turkey is even more prudish in this respect than the common hen. If she once takes to her nest, she will frequent it regularly, even though the egg be removed, as it ought to be as soon as laid.

It should be put in a dry warm place until time to set it. A nest-egg may be left.

When the turkey has laid seven eggs, it is a good plan to set them under a hen, and reserve the remainder for the turkey herself. The hen makes a good mother, and, being lighter, is not so ready to trample the poults as the turkey is; and she can cover well the above-named number, and so bring them all out, as we have found in more instances than one. Should the turkey lay up to eighteen eggs, which she frequently does, then she will have eleven, which is just the proper number for her. Should she fail to lay enough, the number may be made up with hens' eggs. She will take to the chickens as if they were her *very* own, and rear them with the most devoted exclusive attention. So much so was this the case in one instance at least, that when her own poult, hatched by a hen, was added to her brood of chickens, she actually pecked it to death. We have often wondered at the ease with which the feathered tribe were imposed upon in this respect; but no doubt it is a wise provision for the protection of those who have, from any cause whatever, been deprived of true maternal care.

The nest of the turkey may be prepared in the

same manner as that of the hen, but she does not require the same precautions as that fowl; for though at other times disposed to wander, she scarcely moves from her nest during the time of incubation; and so assiduous is she in her duty, that though every egg were removed she would still continue to sit. She may also be placed in close juxtaposition with any other sitter, as she is not jealous of any other being in the apartment with her, like the hen. Food must always be placed near her, or she would be starved to death, so great is her assiduity; indeed, we have always found it imperative to lift her from her eggs, that she might have the needful nourishment. She may have the same food as the hen, and abundance of clean water. The dung and feathers must be removed from the nest every alternate day or so.

The turkey sits four weeks. A day or two before the expiry of that time, the addled eggs must be removed (if there are any), the nest cleared of dung, feathers, etc., and the hen well fed. She must not again be disturbed till the poults have all made their way out of the shell, and are, by the warmth of the mother, fairly dried; if they are, while still wet, exposed to cold, they will certainly die. At the end of twenty-four hours they will be quite dry, and may then be fed. The general practice is to put a pepper-corn down the throat of each; but this may be done or not, as the hen-wife chooses. As to feeding, there are different opinions and plans. We have seen beautiful healthy birds brought up entirely on oatmeal moistened with water or milk, and a little pepper and minced grass mixed in with it. Abundance of clean water was of course provided, and a *small* quantity of milk given daily. This last the poults relished amazingly, and to it their thriving condition was in a great measure to be attributed. Pepper, too, seems beneficial. In a state of nature, turkeys have a great relish for ants, and instinctively discover their nests. These insects emit an acid, for which the pepper may probably act as a substitute. At all events, whatever be the reason, it is almost universally used; and those who do not employ it, admit that it at least does no harm.

Others feed the poults on the same food as that used for chickens, with the addition of hard-boiled eggs for the first week, and at all times minced grass or lettuce mixed in with a small quantity of pepper. Minced nettles also are used; in this case less pepper is required.

In Ireland, again, the principal food is the pickings

G

of porridge at the children's breakfast hour; bits of potatoes at dinner time, or anything that is going; butter milk at all times, and in abundance. The pepper is just thrown down on the porridge, without the trouble of mixing it. If many turkeys are kept by the cottager, of course porridge must be made in greater quantities, perhaps on purpose; and, in that case, the "gudewife" will chop and mix the nettles with it. This *too* abundant weed, being a cheap substitute for pepper, is used to a great extent for this purpose in Ireland. Before being allowed to go out, the poults have a fresh turf brought to them to peck at. In the course of a week after being hatched, they go out and pick the herbage for themselves, and return from time to time to the house, when nature calls for a more substantial diet than that which the field or roadside affords.

The plan to which we formerly alluded as being so successful, consists in giving the poults nothing but hard-boiled eggs—yellow and white minced together; and to continue this until the wing and tail feathers begin to sprout, which will be when they are about three weeks old. For the first fortnight of their lives, poults are subject to diarrhœa, and when it attacks them it is rarely cured. The astringent and nourishing nature of the egg acts as a preventive to this disease, whereas oatmeal porridge, being of a laxative nature, rather promotes it. Water must of course be provided, but milk does not seem to be required till the egg is withdrawn, which the hen-wife will do gradually. She will then give firm oatmeal porridge, with a little milk in the dish to assist the poults in the process of swallowing. Minced cress and mustard will form an agreeable variety; and when at liberty they will pick with avidity the tender leaves of the nettle, and the mother-hen will in due time provide them worms and insects. Such are the different bills of fare: each or any may be adopted, for all are good, and all have been successful; the last mentioned is perhaps the best, but is too expensive for general adoption. It might be modified as follows: Let the sole food be minced egg for the first three or four days, let it be gradually withdrawn, and porridge or oatmeal substituted. Should there be any symptom of diarrhœa, return again to the egg.

Whatever be the feeding, the poults must be kept in-doors for two or three days, after which, if the weather is fine, they may be let out in the heat of the day, to enjoy the fresh air and genial warmth. Should the weather be wet, they must be confined in a house or shed,—indeed for many weeks they must be protected from rain and wind; a thorough wetting is most injurious to them, and often proves fatal. When so unfortunate as to meet with it, they should be brought into the house and dried with a soft towel, placed near a fire, and fed with food in which a little ginger has been mixed. Should the weather be fine, it will be found advisable to put the hen in a coop on the grass, with food and water within her reach. The poults can run out and in at pleasure. Their food must be placed at a distance from the coop, that the hen may not devour it. The drinking-vessel previously mentioned, formed of concentric troughs, ought to be used, for it is of great consequence to keep the poults from getting wetted, which they are sure to do in ordinary saucers or troughs. The food may be placed on a flat plate or board. Like chickens, they must be fed every three hours for several weeks with fresh food continually, to induce them to eat. Should the grass be damp, the coop should be placed on dry gravel, as damp is injurious to every kind of poultry. After this time the hen-wife will gradually remit her attentions, as in the case of chickens, until it will be sufficient to feed them at the same time as the other fowls, and with nearly the same food. But it must not be forgotten that they require stimulating condiments; therefore, if pepper be omitted, nettles, cresses, onions, or mustard should be given. When they are about two months old, the flesny barbles of their head begin to grow, and this is for them a trying season. Some recommend that brine should at this time be mixed with their food, or spirits diluted with water. We would have more faith in an increased attention to warmth, cleanliness, and generous food, to bring them safely through the crisis. Once over this, all danger may be said to be past; they will grow apace, and by Christmas will be ready for market. The demand being at that time great, the sale will be ready, and the prices such as will give the hen-wife a handsome remuneration for her labour. Should she reserve them for her own table, they will, by lessening her poulterer's bill, prove equally profitable. The young cock is generally chosen for roasting, and the hen for boiling. They ought to be in fair condition for the table, if carefully brought up according to any of the plans above mentioned; but, should greater weight be wanted, they can be fattened by any of the methods detailed in Chapter III. of this work. Although proposed in relation to the hen, they are equally suitable for the fattening of the turkey. We must not omit to state that the turkey-cock must be kept apart during the time of the hen's incubation, otherwise he would break the eggs. He will also hurt, or perhaps kill, the young poults if allowed to come near them too soon. The hen-wife will do well to watch the young carefully at their outset in life; she ought not, indeed, to lose sight of them for some time when they first go abroad. So unfit are they to take any care of themselves, that they have not even sense to turn aside from the foot that would crush them—be it man or beast. Their great danger, however, we must confess, arises from the mother-hen herself, who, being careless in her movements, often places her heavy feet on the poor little poults, and, as may be supposed, injures them severely. She is most apt to do this the first few days after they are hatched, probably from her anxiety to protect them; the hen-wife, then, should strive to go as little near them as possible, for fear of frightening the hen.

The plan of hatching proposed above is suitable chiefly where there are only one or two turkeys; but in a farm-steading, where there is abundance of space and a whole flock kept, the method may be different. The eggs may then, if it is desired, be given to turkey-hens, so that the poults may have their natural

mothers, which surely must be the best arrangement
—eleven or thirteen eggs are as many as she can
cover and keep warm. Should any turkey be set on
hens' eggs, twenty will be found sufficient. The first
trying weeks of the poult's existence over, it will
luxuriate in the stack-yard and stables; but if the
farmer does not look well to his fences, it will make
sad havoc in his fields of standing corn. In autumn,
when they are turned into the stubble with the other
fowls, they will gain rapidly both flesh and fat.
Should the farmer use his light grain in the fattening
of large flocks of turkeys, he would find it much
more lucrative than parting with it at the low price
it would bring in the market. If these creatures pay
when reared by the private gentleman,—as pay they
do, and that handsomely,—much more must they pay
when consuming only what would be undersold on
the one hand, or go to waste on the other. Before
concluding the subject, we must not forget to warn
those who have a turkey-cock about their premises,
of the danger children incur by being in its neighbourhood. The turkey is fond of bread, and should
he see a child with any in his hand, he will run after
him and snatch it away: should the child resist, as
it is quite likely he may, the cock may get furious,

and injure the child seriously. And one peculiarity it has, and it is this: if any one runs from it, it will follow with the bravery of a lion; if, on the other hand, you boldly chase him, he will flee like the veriest coward. Hence the danger to children, who are so frequently running.

The next in our list is the goose. And the cottager, whose abode is in the neighbourhood of a stream and a common, could not possibly fix on more profitable live stock. Even the roadside and the pond will afford them subsistence. Let out in the morning, they will forage for themselves during the day, and return at night in want of nothing but shelter. Surely this is " profitable poultry." We grant, however, that it cannot always be so to this extent; they must in general be fed, but they are content with little. Grass and water are what they chiefly require; this, with a fair supply of oats, makes up their bill of fare. The goose commences to lay early in the season. Before doing so she may be seen sitting among the straw, and picking and carrying it as if to make a nest. When this is observed, a nest must be prepared for her, in which, if she takes to it, she will lay an egg each alternate day. When she has laid about twelve or fifteen, she will show a desire to sit, and eggs should then be given, twelve or fifteen, according to the size of the goose. Water and oats are placed beside her, but she will not eat much while she sits. A little boiled potato may be given her occasionally to prevent constipation. She need not be hindered going to the water when she pleases to do so; it will refresh her and make her sit all the better, and will not in any way injure the eggs. The hen-wife need not, as in the case of the turkey-cock, keep the male apart, for the gander takes the greatest interest in the process of incubation, and keeps the most affectionate watch over his mate. While the goslings are being hatched, the goose must be watched, but not disturbed. After the goslings are all out, which will be at the end of a month, they may be removed, even before being dried, to the heat of a fire: some even recommend them to be taken to a dry spot in a grass field for a few hours in the heat of the day. The sun's rays, it appears, dry them much more speedily than the warmth of the parent-bird. This is, however, rather a hazardous plan, as, should the weather suddenly change, and the goslings get wetted during even the first two days of their life, they will inevitably die. As the goslings do not all make their appearance at the same time, the goose is apt to march off with those first hatched, and leave the rest to their fate. To prevent this, it is well to remove them, one by one, as they make their appearance; they can be put into a basket, lined with flannel or wadding, by the side of a fire or in the sun, and returned to the goose when they are all fairly hatched. And we may here take the opportunity of saying that this may be done with the young of all domesticated fowls, as well as with the goose, when there is any irregularity in the time of hatching. With the young of the turkey, we may add, particular care must be taken, as they will be sure to die if exposed to cold at this early stage of their existence. But to return to the goslings. The best food at first is gruel made of barley or oatmeal —stir-about, as the Irish peasantry call it — they gobble it up like ducks. Water must be given them in shallow vessels, so placed as not to be overturned. The best plan is to have the concentric troughed saucer previously mentioned; it is of consequence to keep the goslings from being wetted at first. Should the weather be such that they cannot go out, a fresh turf must be brought into their house; but should the weather be fine and sunny, they will do best in the open air, on a dry grass field. They must, however, even then, be frequently fed; for, if at first limited to grass, they will, in all probability, take diarrhœa and cramp, lose their strength, and pine and die. A watchful eye must be kept over them for a few days, for if they fall upon their back, they cannot regain their feet, and if unobserved they will be sure to die. Should .there be any small pond in the neighbourhood, in a retired place, secure from the presence of horses, or other large animals that might be dangerous, the goslings might go with the mother-bird to it, and venture to swim. In a few days more they will be past all danger, and in a month will be found to have grown amazingly. They will then require nothing more than a feed of oats along with the larger birds before being turned out into the fields in the morning. On their return at night they must again be fed with oats. This must, on no account, be forgotten, as it induces them to return with punctuality.

When many geese are kept, it is usual to entrust the care of them to a boy, called a goose-herd. He leads them out in the morning to the fields and to the pond, and brings them back in the evening. The grass field in which geese herd must be reserved specially for them, as their droppings prevent cattle or sheep grazing upon it. But we suspect the same result would follow with any other species of fowl. With the above-mentioned treatment,—oats, Indian corn, or any kind of grain, morning and evening, good pasturage during the day, with access to a pond in which they find the animal food which suits them, —by the approach of harvest they will have attained size and flesh to a considerable extent. Harvest over, they can be turned into the stubble, when they will soon gain sufficient fat for ordinary tastes. If additional fatness be wanted, they can be fattened on oatmeal and milk, in a similar manner to that recommended for fowls and turkeys in Chapter III. of this work.

Some farmers think it a profitable plan to purchase geese when young and lean, and turn them into the stubble to fatten. When fat, they pluck and resell them. This is practised in some parts of the Continent. The profits form part of the daughters' dowry—"dot." Others, again, buy them earlier in the season, pen them up, and feed them on the garden-produce, then so abundant—cabbage, lettuce, turnip, and beet. To this they add dry oats, and abundance of clean water *to drink*. They get sufficiently fat by this treatment in a fortnight or so, and their flesh has a delicate flavour, greatly preferable to the rank taste of the over-fed and crammed goose. It may here be stated that cabbage, lettuce, turnip-tops, etc., may very properly be added to the bill of fare which we have proposed for the young gosling. Indeed, for all kinds of poultry, the three kinds of food, *soft*, *dry*, and *green*, are highly suitable, nay, necessary. As to the proportion in which they require to be kept, five geese will be found sufficient for one gander.

Not only are geese valuable for their flesh—their down and feathers are useful to the housewife, forming cushions, pillows, and beds. The quills procured from their wings were formerly of great value, but the introduction of the metallic pen has somewhat lessened it. The feathers are plucked three times a-year—in July, in October, and in December. The feathers of lean geese are better than those of fat ones, and those of living geese better than those of dead ones. The feathers are plucked in July from the young geese, and the down from under the belly, wings, and neck. The feathers, though from a living goose, not being come to maturity, are inferior to those plucked in October, which, being taken at the season when they would naturally fall off, are in per-

fection. The feathers, again, in December, bein
taken from a dead fowl, are inferior to those pro
cured in July. On this principle, it is evident tha
the sooner the goose is plucked after it is killed th
better.

As to the varieties, the common grey or mottle
goose is the usual stock in this country. The gan
ders are generally white, and the geese are grey
mixed with dull brown; sometimes, however, the
are also pure white. In this case they are ver
beautiful sailing on a river, or on a sheet of watei
they are like so many swans.

The Bremen or Embden goose is a beautifu
variety. Its size is large, and its plumage is a spot
less white. They are of a quiet domestic characte:
do not wander from their pond, and fatten easilj
Their feathers are very valuable. One objection i
their laying too early for this cold northern climat
They sit and hatch well, and frequently bring out tw
broods in a season.

A curious and beautiful variety, exhibited at th
Crystal Palace, London, deserves notice. It is cha
racterised by a curly plumage, which gives a peculia
look, as if it were quite a distinct species. To quot
from an agricultural work,—" The feathers on th

back are curled and frilled upward; the secondary, or smaller feathers of the wings, are elongated and twisted. The tail coverts have also the same peculiarity. Their habits are precisely similar to those of the common goose. But one pair have been taken to England, and we have not learned what is their value for the table." Perhaps the curly nature of the feathers makes them more elastic and better for beds. At present they are regarded only as " fancy fowls," for gratifying the curiosity of those interested in poultry breeding. The pair of geese above mentioned were brought from Sebastopol.

Far from being stupid, as is generally supposed, the goose is possessed both of intelligence and affection. We know of one which formed an extraordinary attachment to an old gentleman, an attachment which she lost no opportunity of testifying, according to her own fashion. She was in the habit of watching daily for his appearance, and the moment he was within sight she spread her wings and flew towards him, evincing the greatest joy and delight in his presence, and following him to the door of his place of business. On Sundays she would accompany him to the church-gate, and on the dismissal of the congregation would be found patiently waiting for the appearance of her favourite, whom she would again follow to his residence. Unfortunately her antipathies were equally strong, for she took a dislike to another individual in the same town, and showed that feeling in a manner so marked, and at the same time so disagreeable, if not dangerous, that complaints were made to her owner, who was thus reluctantly compelled to part with the offending animal. She was therefore sent into the country, where, far from both her favourite and her foe, she lived to a good old age. This fact, though strange, is perfectly true, and can be authenticated as to place, name, and date.

The duck next comes under our notice. Of all the different varieties, the Aylesbury is the most esteemed; its colour is a spotless white, it is a good layer, but the eggs are smaller than those of the Rouen duck. The plumage of the wild-duck is reproduced in that of the Rouen duck. The points of the common duck it is as impossible to specify as it would be to do that of the common barn-door fowl.

The duck is much esteemed as a table-bird; though inferior to the goose in size, it is thought by many to be superior in delicacy of flavour; its eggs are not so much liked in general as those of the hen, but they are much used in cookery; their feathers are very valuable, and are only inferior to those of the goose. The duck is not at all fastidious as to food; will gobble up any refuse from table, kitchen, dairy, or garden—nothing comes amiss to it. It is very useful, too, in clearing the garden of slugs, snails, etc., and does not injure plants and flowers like the common hen. Some keep ducks where there is no water, and find them to thrive and fatten; but this is a cruel plan, for water answers the same purpose for them as does the dust-bath for the common fowl. If deprived of their natural element, which is the only mode of cleansing open to them, they will become infested with insects, which will not only torment them exceedingly, but prevent their thriving. Water is even more necessary for ducks than for geese; it is miserable to see them dabbling in the dirtiest, filthiest puddles. Little wonder that their flesh, in such cases, should be coarse, and their eggs strong and rank. When there is no stream or pond, a tub should be sunk in the ground, filled with *pond* water, in which they may wash and refresh themselves. Their food may be the same as that previously recommended for hens, only instead of being friable, may be rather wet, should they feed with the hens. If a drinking-trough be near the food it will assist them to swallow it even in its

H

friable state. Green food of all kinds, either boiled or raw, or both, will be found suitable; boiled barley will be found a good variety for them. They do not roost like hens, but sit on the ground. The floor of their house should be brick, to allow of frequent washing; it should be littered with straw, and the straw changed every alternate day. This floor should be frequently washed. They may be lodged along with the geese, while the turkeys and hens can roost together.

The duck begins to lay early in spring. If not well looked after she will deposit her eggs in the fields or yards, or even the ponds, where they will be lost to the hen-wife, and only serve as food to the pigs or to the frogs. A comfortable nest must be placed for her in her house, and she ought not to be let out in the morning till she has laid. Owing to the difficulty there is in getting her to sit in any nest but the one in which she is accustomed to lay, it is usual to employ hens to hatch her broods. She is, besides, not a very good mother, and is apt to take the ducklings too soon to the water; for, notwithstanding their aquatic propensities, ducklings, like goslings, will die if wetted during the first days of their existence. It will be necessary, then, to coop up the mother-bird, whether hen or duck, for the first fortnight. At the end of that time the ducklings may go to the pond, provided they are in no danger from larger animals. They are very senseless as to danger, and will allow themselves to be trodden down, without any effort to save themselves; attention, therefore, must be paid to them for some weeks. The hen will evince laudable anxiety in seeing her adopted ones enter the water; she will, however, receive them again with fondness, and tend them with maternal care, unconscious of the imposition practised upon her. The food may be the same as that used for geese: thick gruel of barley-meal; cold porridge is also very good. In a few days boiled potatoes may be given, and green food must not be forgotten. Water must be supplied in such a way that the ducklings will not overturn the vessel, for they must on no account get wetted at first. When they have been confined to the coop for a fortnight or three weeks, they may, as previously directed, be allowed more liberty; and should a small retired piece of water be in the vicinity, it need not be forbidden them. They will now give the hen-wife very little trouble. If fed in the morning, and turned into the field among the geese, they will forage for themselves during the day; and an evening feed will induce their regular return to their night-quarters, which must be clean and comfortable. If housed among the hens, they must not be placed under the roosts, but straw must be placed for them in a retired corner. This straw must be frequently changed, or it will soon get wet and dirty. A warm and dry bed is as needful for them as is warm housing for the hen.

Peas and beans form an excellent addition to the food of the duck; it gives firmness to the flesh. Barley-meal is said to make it tasteless and insipid. This can only be the case when it forms the sole food, for when combined with others it is excellent. Bran, sharps, or middlings, may be largely used with ducks, when combined with potatoes, green vegetables, and pot liquor. Oatmeal is confessedly one of the most nourishing kinds of food for poultry; but, being more expensive than many others, we have only recommended it for the young in the poultry-yard, and for the adult in the fatting-coop. Should the duck be allowed the run of the barn-yard, and then, with the turkeys and geese, be turned into the stubble, she will luxuriate and fatten amazingly, sufficiently so for any ordinary taste; but if more is required, she can be shut up into the fatting-coop, and be treated as has been already directed. The appetite of the duck is at all times keen, the profit to its owner being thereby lessened; yet, if near a pond, a roadside, or a common, it might be kept so as to bring a fair return for the outlay; and the cottager who finds himself in such a situation should never be without half a dozen waddling about, for to him they would come under the denomination of " profitable poultry."

Before leaving this subject, it may be as well to state that, to a great extent, the same plans recommended for the management of the hen, may be followed with regard to the fowls of which this Chapter treats. If a cockerel kept in the yard after it is ripe will " eat its own head off," much more will the turkey, the goose, and the duck do .so, and that, too, in a much shorter period. Let these fowls, then, be disposed of in succession the moment they are ready. Keep them always young, and they, too, will be equally profitable as the fowls. On the 25th December the demand will be immense, the profit will be great; for who would grudge a price that wishes to grace his board on that joyous day with a Christmas *goose?* No other dish would make up for the want of this; its time-honoured presence seems to make the family gatherings even still more familiar.

The young stock by this time will be despatched. A few may be reserved to breed from in the spring; a gander and six geese, less or more, according to what is wanted. The field in which they have been fed,—by this time sufficiently bare and sufficiently soiled,—will now lie fallow, so to speak; and when the goslings are hatched and ready to be turned out in the spring, will be again in fine condition, and the farmer will have no cause to regret the use to which he has put his acres, at least if the hatching has been successful. And here let us state, that if the goose has not food and water placed beside her, adieu to all hopes of a brood, for she will, to a certainty, eat every egg. The hen-wife, then, will attend to this, and supply the patient goose with the little that she needs.

As has been already mentioned, should the geese and ducks lodge in a corner of the hen-house, they must be supplied with a warm bed of clean straw. They should be kept in the house till after they have laid. To prevent them getting out, the small hole by which the hens enter should be made some feet from the ground; access to it may be by a small ladder. Besides preventing the ducks and geese from getting out, it will be a protection to all from the attacks of rats, pole-cats, etc. A fowl-house filled with occupants so varied is a pleasant sight to the lover of nature. Some have asserted that such a variety ought not to be together, that the larger fowl should be separated from the smaller, the aquatic fowl from the barn-door hen; but, for our part, we have seen all agree well in company. At first, a little bickering might be observed, but a short acquaintanceship was sure to establish harmony and peace. Nay, we have even seen the gentle dove dwelling unmolested in the midst of this mixed assembly. A solitary pigeon had lost its way, and found a resting-place in a hospitable homestead. At first it kept aloof, sitting in the sheds, or on sheltered window-sills. Becoming less shy, it drew a little nearer, and passing by the barn-door hens, which might seem to be its nearest relatives, it began to make acquaintanceship with the ducks, followed them, and at last fed with them. Then it accompanied its new friends at night to the hen-house door, and in the morning would be seen watching and waiting with the utmost anxiety for their exit. Joining company with them, it would strut in their midst while they waddled to the pond. There it would hover about, delighted at every return of its friends to *terra firma*, strutting constantly be-

side them, and only leaving them when they entered
the hen-house. At last, instead of merely accompany-
ing them hither,. it entered with them, and fairly took
up its abode in more comfortable quarters than it had
at first chosen. The kind hostess, however, pitying
it unnatural condition, procured companions for it of
its own kind. And, conceiving that it would be more
comfortable in a "house of its own," a dove-cot was
also provided. This latter arrangement, however, did
not suit Madam Pigeon's taste; one night's occupancy
of her "own house" was sufficient. She returned to
the hen-house, introduced the new-comers to her
former associates, among whom they also took up
their abode. Two tiny eggs were soon laid in a hen's
nest; in due time two young pigeons appeared. Unfortunately one died, perhaps crushed by an intruding
hen; the other survives, fed by the parent birds,
without any care on the part of the hen-wife, and
bids fair to be seen, ere long, strutting by the side of
the waddling ducks. We are glad to say that there
is now greater hope of this happy result, as, while the
above was being written, the kind-hearted host and
hostess of the hospitable homestead had caused a neat
little dove-cot to be placed *in* the hen-house. Being
at some height on the wall, it will preserve the young

from the intrusion of hens, and also of other enemies which might attack them. Their danger being thus lessened, the hope of rearing them will be increased; and we are happy to say that the old pigeons have at once taken to their new abode. And truly it is an interesting sight, both to the hen-wife and to the lover of nature, to see, under one roof, such a varied assemblage of the feathered tribe,—the common hen, the duck, the goose, the turkey, and the pigeons. Nor does this complete the list of the individual inmates in this happy hen-house. Years before, a pea-hen had, like the pigeon, found its way to this same homestead—whence it came was never known, but hearty was the welcome given, and kind the treatment it received. Its gentleness made it a favourite with all, and its elegant appearance made it quite an ornament to the shrubbery, as well as to the poultry-yard. A mate in this case could not be had; and, unlike the pigeon, the pea-hen formed no friendships. Proudly she towered above all her feathered associates, as if the aristocrat could not stoop to the plebeian. In peace, however, she dwelt among them. If any presumed to snatch with vulgar haste the morsel which she with such slow majesty was preparing to take, she might possibly attack them with fierceness, but generally she passed over the insult, as if the insulter were beneath her notice.

According to the instinct of her species, she would, at the approach of spring, wander into some retired part of the field, deposit her eggs, and then sit upon them. When hunger pressed her, she would appear at the hospitable door, receive the ever-ready supply, visit the water-trough, and return again to her weary work—weary and hopeless indeed; for, it need scarcely be added, no little ones ever appeared. In order to gratify her keen maternal instincts, hens' eggs were given her; but, whether her weight was too great, or the warmth of her body too much, we know not, but certain it is, no living chicken ever appeared; so the poor affectionate creature was doomed to perpetual disappointment.

Although the pea-fowl does not come under the denomination of profitable poultry, yet, as it has incidentally come under our notice, we cannot refrain from continuing the subject, and giving a few hints as to its nature, and some directions as to its arrangement. The male bird is of surpassing beauty; its gorgeous hues can scarcely be described, but must be admired by all. It forms a splendid embellishment to the lawn and to the shrubbery, although its hoarse

and screaming voice is to some a great drawback.
The female is not destitute of beauty, although far
inferior to the male. She has a pretty tuft on her
head, but there is no trace of the star-bespangled tail
which forms such an ornament to her mate. She is
of a timid and gentle nature. When the laying season
commences, she seeks a sequestered spot, which she
carefully conceals from the peacock, who would be
sure to break the eggs were he to discover it. Should
the nest be found out, the hen will be sure to forsake
it; it must, then, be cautiously approached, and the
eggs stealthily removed, one or two being left as a
decoy. The others must be kept in a warm place,
and be cautiously returned to the nest when the hen
has finished laying. The pea-hen will by no means
be restricted as to situation; she will sit in the place
which she chooses, and nowhere else. We have heard
of one which disappeared from her owner; months
after, it was discovered that she had wandered into the
adjoining woods, laid her eggs, and hatched and reared
her chickens. One visit she paid to her home, accompanied by her thriving family, but every effort to
secure them failed; being fully fledged, they took
wing back again to their native woods, where they
now rejoice in liberty unrestrained. How interesting

it would be to meet these splendid birds in the forests of this northern isle: one might fancy themselves transported into the bright and sunny climes of which this bird is a native. But to return to the subject. The eggs may, if desired, be given to a common hen to hatch. When hatched, the young may be fed much in the same manner as the turkey-poults; and when a month or two old, can be fed on boiled barley, or other grain, along with the parent bird. The hen, if the natural mother, must be confined in a coop, otherwise she might make off to the woods with her treasures. She knows full well the danger they incur from her mate, and therefore shuns him by all means, and keeps as far as possible from his precincts. How interesting to trace the guidance of instinct in one and all of our feathered favourites!

The pigeon, too, having the credit of consuming untold quantities of grain, scarcely deserves to come under our notice, our plan having reference only to stock that will pay. Still, having introduced the pigeon, we must devote a short space to its management. And, in the first place, we would say that the treatment recommended for poultry of all kinds must be adopted here, viz., to keep them always young. Feed them well until they have size and flesh sufficient, but let them not have a grain after that. This plan pursued, they, too, will be found profitable. Unlike the occupants of the poultry-yard, pigeons associate in pairs. The female commences to lay when she is about nine months old. She lays two eggs in three days, sitting partially during this time. She then, assisted by her mate, sits for fifteen days, at the end of which period the chicks appear. They consist generally of a male and female, and are fed by the parent birds for the first few weeks. When they have attained some size, and are just about to take wing, they should be killed for the table, and the adult birds reserved for breeding. The interior of the pigeon-house may be fitted up with shelves, about eighteen inches apart, divided into small compartments for nests. Before each nest may be placed a slip of wood, on which the pigeons may coo. It is not necessary to furnish the nests, as in the case of hens, for the pigeons carry sticks and straws for themselves. The floor of the house should be strewed with gravel, it being as necessary for the health of pigeons as of fowls. There should also be water provided, not only for drinking, but also for the purpose of ablution. The pigeon is very prolific; it has been calculated that one pair will produce in four years

upwards of fourteen thousand birds.   Presuming
that this calculation is correct, we cannot but think
that such an astonishing increase indicates that
they were designed for the use of man.   In view of
this prodigious increase, some assert that vast profits
may be made by these birds.   Others, again, affirm
that they are quite ruinous, and counsel the farmer
to have nothing to do with such a voracious tribe.
Perhaps the truth lies mid-way.   Certainly, if well
and regularly fed, they cannot commit great ravages
on the standing crops; and the young pairs, if successfully reared, will bring prices affording a fair
profit for the food of the old ones.   They will feed
nicely with the fowls in the poultry-yard, picking
even the soft food we have recommended.   Oats,
barley, and, indeed, all kinds of grain, are their delight.   Small horse-beans are their favourite food,
and should be liberally supplied; also peas, either
grey or white.   Hemp and canary seed should also
be given occasionally.   Salt should be sprinkled
about their cot.   The dove-cot must be frequently
cleaned.   The dung accumulates rapidly, and will
amply repay the trouble of removing it.   It will be
found very valuable for agricultural purposes, and is
an important item in calculating the profits.   And

not only the dove-cot; the poultry-yard also will contribute largely to the manure depôt, and in any calculation of profit and loss must not be forgotten

The care of pigeons is a favourite occupation for boys. It is also a useful one, for it employs time which might otherwise be spent in idleness or mischief. It also tends to cherish in their minds a kind and loving disposition; for the roughest nature, if brought into frequent contact with the gentle dove, must in time have its asperities smoothed down. It will be found an excellent method of promoting in the young that consideration and thoughtful care of others, which every right-minded parent is anxious to see in his offspring, and which the best meant admonitions often fail to produce. One drawback, doubtless, is the putting to death of these lovely creatures. In the present state of things, we cannot see how this can be avoided. In the glorious future, which cannot now be far distant, it will be different. The animal creation will then, we have reason to believe, be delivered from the evils under which they now groan. At present, our part is to add all we can to the happiness of their short existence, and in mercy to make their death as easy as possible. Their happiness is best promoted by leaving them, as much as circumstances permit, to follow out the instincts of their nature. The process of incubation, which seems so wearisome to us, to them is a source of the most intense pleasure; and the rearing and tending of the young is, as we can easily suppose, a task even more delightful. But when the young are old enough to forage for themselves, the care of the mother ceases, and to deprive her of them at this period gives her little or no uneasiness. We have often noticed, in the case of the domestic hen, that notwithstanding the affectionate solicitude which she in general evinces towards her young, should any of them be suffering from disease, she will neglect it, nay, even ill-use it, and leave it to perish; and should any of her brood wander from her, she will gather the remainder under her wing, and not even miss the absent one. It is probable, then, that the pain endured by the parent pigeons on the removal of their full-fledged offspring is both slight in intensity and short in duration; and if no wanton cruelty be used towards the young, the pang they suffer in parting with life will be but momentary.

CHAPTER VIII.

CONCLUDING REMARKS.

WE have now conducted our readers through the successive stages of the hen-wife's journey, and feel assured that, even though previously utterly ignorant of the road, she is now prepared to tread it with safety and with success. In other words, we are assured that the above directions will be found sufficient for every practical purpose, as far at least as regards useful and profitable poultry. The ornamental and fanciful we leave to others, fully content to attain the humbler object which we proposed to ourselves, and announced to our readers, at the commencement of this work,—better pleased still if our twofold object be attained, and our readers be persuaded to join heart and hand to promote the scheme which we have been advocating. The mighty motive which we urged at the outset still exists, and exists with undiminished strength. The frightful disease

which was then advancing through our land, instead
of retreating, keeps. its place; nay, proceeds, and
gathers strength as it proceeds.   The hope which
then existed that vaccination might prove a protection, or at least might mitigate the severity of the
disease, has now vanished away, and with it also
every hope of a cure.   It becomes, then, our duty to
use every means to prepare for the scarcity which must
soon supervene ; and one way assuredly is, to increase
every kind of stock not affected with the fell disease.
The claims of the Poultry-yard it is our part to
advocate; and we would urge upon our readers to
begin, without a moment's delay, to hatch, to rear, to
push poultry into the market; where, by the time
they are ready, the demand, it is to be feared, will
only be too great, and the prices consequently pro-
portionably high.

We trust our readers will at once enter into the
scheme earnestly and vigorously, and success, we
doubt not, will speedily follow.   The quick rate at
which the domestic fowl multiplies, as compared
with the wild species, is a strong proof that it
is intended for the use of man, and gives a great
encouragement to the right-minded worker to per-
severe.   Let the hen-wife, then, listen to the first

welcome sound that indicates the desire to brood. She will thus have the pleasure of counting up, in due time, her chickens by scores, or even by hundreds; and also the delight of gratifying the strong instincts of the mother-bird, which to a humane mind is no slight consideration. So strong is that instinct that the hen, if left at liberty, will gratify it at all hazards. We know of one which burst through all restraint, and so far returned to her original habits as to make her nest in the branches of a tree, and there to hatch and rear her young, bringing them in triumph to the barn-yard. Another, undeterred by the rigours of a northern winter, sought and found a nest for herself—where, it was never known; but snug and warm it must doubtless have been, for thirteen chickens were reared in it; and with this noble brood the mother-hen walked through the farm-steading into her accustomed quarters, and that on Candlemas-day, so that her brood must have been hatched in mid-winter. Warm, indeed, was the welcome given them by the good lady of the homestead, and comfortable were the quarters she provided. In the very same homestead, in a different year, but on that very same day (Candlemas), another hen also walked through the yard with her young brood, but she,

poor thing, was less fortunate; her nest in the fields
was less warm and snug; rheumatism was the result,
and she remained lame for the rest of her days. If,
in such untoward circumstances, the hen braved the
blast, and survived to bring forth her brood, be-
numbed and paralysed as she was, the hen-wife need
not be afraid, with her superior accommodation, to
hatch and rear at any season, however far advanced.
Greater care will, no doubt, be necessary, and fewer
chickens will be reared; but those few are valuable,
coming in, as they do, when fowls are rare, and consequently high-priced.

The tenacity with which the domestic hen adheres
to life is truly surprising; cases are on record which,
were they not well authenticated, we would pro-
nounce incredible. We know of one, where a hen
had a cart-load of straw emptied upon her, and which
was found alive when the last of the straw was removed, a full month afterwards. The only food to
which she had access during this long period was her
own produce, which, we need scarcely say, she had
eaten.   And, in passing, we would say, that, as a
general rule, hens, if properly fed, will not eat their
own eggs. We can scarcely blame them for doing so
when pressed with hunger. Another case still more

wonderful, and we have done. A country gentleman had a hen which presumed to cluck when she was not wanted to do so. Without calling for the assistance of any one, he, with his own hands, placed a tub on the disobedient hen. Business called him suddenly from home, and he forgot all about her. The tub did not happen to be required, so the poor hen was not missed, and for weeks remained in her living grave. At last the gentleman returned, but not to set the poor prisoner at liberty, for the circumstance had entirely escaped his memory. At the end of *six weeks*, however, the thing suddenly flashed on his mind. We need not say how rapidly he rushed to the spot; we need not describe his surprise and delight on finding the neglected hen alive,—stupefied and giddy on being restored to the light from which she had been so long shut out; weak too, but with strength sufficient to totter to the adjoining stream and to drink abundantly of its waters. This gave her new life; and though for some time weak, she was ultimately restored to her former condition. During the time of her dreary confinement, she had scraped all round the spot to which she had access, and doubtless must have found a few scattered worms in the ground; but, at the most, the nourishment during that long period must have been scanty indeed, and were the case not fully authenticated, might be deemed incredible. Besides showing the extreme tenacity of life in fowls, it proves what we have insisted so much upon—the necessity of a copious supply of water. The exclamation, "Who ever heard of giving water to hens?" could only be uttered by one totally ignorant of their habits and wants.

In a former part of this work, we have urged the advantage of devoting some portion of pasture-ground to the rearing of the larger kinds of poultry—turkeys and geese,—the latter especially; and now that every day is adding strength to the motives for doing so, we would urge it with still greater earnestness. The destroying scourge, which has emptied so many stalls and thinned so many fields, must have left unnumbered acres unoccupied, free to be employed in the manner we advocate. Even in districts unvisited by the destroying plague, pasturage is not wanting for the same purpose. In numerous cases, the farmer, rather than risk his cattle, has sent them all, fat and lean together, prematurely into the market; while in others, although the hope of high prices may have led him to risk the keeping of those already in possession, yet the gloomy aspect of affairs, as well as the restrictions imposed, makes him chary of adding to their number. Thus a large breadth of pasture-ground must be free to be employed as the farmer may choose. He may indeed convert it all into corn-fields; but we conceive that he would find it a gain to reserve some portion at least, and that no inconsiderable one, for the purpose we propose. The turnips, too, which are in general consumed ere this, being left for want of consumers, will be found useful. If cut down in slices, the geese will greedily devour them, and will, in consequence, require less corn. If thrown whole into the poultry-yard, the fowls will pick out the inside, and leave only the outside shell. Beet-root and mangold-wurzle are also highly relished by them; and at an advanced season of the year, when the green crops are consumed, will be found very valuable. Turkeys, when reared, are very hardy, and will bring a large profit to the rearer. Geese are even more easily reared, and still more easily fed. Their flesh, though perhaps deficient in delicacy, is more suited to the strong, the robust, and the healthy, than is the more delicate flesh of the domestic fowl. The down and fine feathers which they produce so abundantly add greatly to their value. The duck may be associated in the field with the goose. It will not bring so much profit to its owner, being a voracious feeder; its down and feathers, however, are valuable, and its flesh, though somewhat similar in flavour to that of the goose, is esteemed more delicate.

In fine, we can only express a hope that the arguments we have adduced may commend themselves to the judgments of our readers, and that they will adopt, without delay, the line of action indicated. In the farm-steading, at least, there need be no difficulty; there, all that is needed for carrying out the plan on a sufficiently extended scale is to be had— housing for fowls of every variety, food of every kind, and space abundant in which all may roam. A great good will thus be accomplished—our food-supplies will be materially increased—our poultry will soon rival those of our Continental neighbours, — and those supplies which we have hitherto so largely received from them will, it is to be hoped, eventually be rendered altogether unnecessary.

Schenck & M'Farlane, Printers, Edinburgh.